SUSTAINABLE AGRICULTURE IN CENTRAL AMERICA

Sustainable Agriculture in Central America

Edited by

Jan P. de Groot
Department of Development Economics
Free University of Amsterdam
The Netherlands

and

Ruerd Ruben
Department of Development Economics
Wageningen Agricultural University
The Netherlands

First published in Great Britain 1997 by
MACMILLAN PRESS LTD
Houndmills, Basingstoke, Hampshire RG21 6XS and London
Companies and representatives throughout the world

A catalogue record for this book is available from the British Library.

ISBN 0–333–68228–9

First published in the United States of America 1997 by
ST. MARTIN'S PRESS, INC.,
Scholarly and Reference Division,
175 Fifth Avenue, New York, N.Y. 10010

ISBN 0–312–17555–8

Library of Congress Cataloging-in-Publication Data
Sustainable agriculture in Central America / edited by Jan de Groot
and Ruerd Ruben.
p. cm.
Papers from the 1995 annual conference of the Association for
European Research on Central America and the Caribbean, held Oct.
14–15, 1995, in Paris, France.
Includes bibliographical references (p.   ) and index.
ISBN 0–312–17555–8 (cloth)
1. Agriculture—Economic aspects—Central America—Congresses.
2. Sustainable agriculture—Central America—Congresses. I. Groot,
J. P. II. Ruben, Ruerd. III. Association for European Research on
Central America and the Caribbean. Conference (11th : 1995 : Paris,
France)
HD1797.S87 1997
338.1'09728—dc21   97–9144
                                                                      CIP

This book is printed on paper suitable for recycling and made from fully managed and
sustained forest sources.

10   9   8   7   6   5   4   3   2   1
06   05   04   03   02   01   00   99   98   97

Printed and bound in Great Britain by
Antony Rowe Ltd, Chippenham, Wiltshire

Contents

Notes on the Contributors

Rosario Ambrogui has a master's degree in environmental economics from the Universidad Nacional (UNA) of Costa Rica; she is a lecturer at the Escuela de Economía Agrícola of the Universidad Nacional Autónoma de Nicaragua, ESECA-UNAN, Nicaragua.

Johan Bastiaenen is senior lecturer and researcher at the Centre for Development Studies of the University of Antwerp (UFSIA), Belgium; he has published on peasant production systems and the development of non-conventional sustainable rural finance, in particular in Nicaragua, working with NITLAPAN, the research centre for rural development of the Universidad Centroamericana, UCA.

Marrit van der Berg is PhD candidate, working at the Department of Development Economics at Wageningen Agricultural University, Wageningen, the Netherlands; she did fieldwork on land evaluation in agricultural production cooperatives in Honduras.

Harry Clemens works at the Centro de Desarrollo Rural of the Free University, Amsterdam, the Netherlands (CDR-VU) in San José, Costa Rica; he has research experience in Panama and Nicaragua, where he worked with the Escuela de Economía Agrícola of the Universidad Nacional Autónoma de Nicaragua, ESECA-UNAN.

Jan P. de Groot is senior lecturer and researcher in rural development at the Faculty of Economics, Free University, Amsterdam, The Netherlands, with a long-time experience in Latin and Central America; he has published on land reform and sustainable agriculture, in particular in the humid tropics.

Sarah Howard does her PhD research at the School of Earth Sciences of the University of Greenwich; her field research focuses on conflicts over land and resources in the Bosawas area in Nicaragua.

Kees Jansen is PhD researcher at the Department of Rural Sociology of Wageningen Agricultural University, Wageningen, The Netherlands; for his dissertation he did fieldwork in a number of rural communities in the north of Honduras.

David Kaimowitz was for many years a specialist in the area of science and technology, natural resources and agrarian production at the central office of the Instituto Inter-americano de Cooperación para la Agricultura, IICA, San José, Costa Rica; he now works at CIFOR, Jakarta, Indonesia.

Mario Lopez has a master's degree in rural development of the Institute of Social Sciewnces, ISS, the Hague, The Netherlands; he is lecturer at the Escuela de Economía Agricola of the Universidad Nacional Autónoma de Nicaragua, ESECA-UNAN, Nicaragua.

Ruerd Ruben is senior lecturer and researcher in agricultural economics at the Department of Development Economics at Wageningen Agricultural University, Wageningen, The Netherlands; he has a long-time experience in research and policy formulation in Central America and has published on topics such as land reform, production cooperatives, sustainable agriculture and institutional economics.

Andy Thorpe is senior lecturer and researcher in development economics at the School of Economics, University of Porthmouth, England; he was visiting professor in agricultural economics at the Posgrado en Economía y Planificación de Desarrollo (POSCAE) de la Universidad Autónoma Nacional de Honduras, UNAH.

Peter Utting works at the United Nations Research Institute for Social Development (UNRISD), Geneva, Switzerland; he has published extensively on the social and political dimensions of resource degradation, in particular in Central America.

Daniel Wachter, until recently senior lecturer and researcher at the Geographic Institute of the University of Zurich, Switzerland, is now working at the Swiss Federal Office of Spatial Planning; he has been engaged among others in research on property rights and land titling in relation to sustainability of agrarian production systems.

Cor Wattel is coordinator of the Centro de Desarrollo Rural of the Free University Amsterdam, the Netherlands (CDR-VU) in San José, Costa Rica; his research in several Central American countries is in particular related to rural finance and institutional development.

1 Introduction and Summary
Jan P. de Groot and Ruerd Ruben

1 INTRODUCTION

Processes of deforestation, erosion and resource depletion have been particularly severe in Central America during the last fifty years. Agricultural development in this region has been based mainly on extensive growth, supported by macroeconomic policies oriented towards the promotion of agricultural exports. Ongoing urbanization has been accompanied by peasant producers becoming involved in commercially oriented agriculture, and by intensifying pressure to increase the productivity of land and labour. However, access to improved inputs (seed, fertilizers, pesticides) has been limited to a small segment of middle class producers.

In recent years, public and private interest in the sustainable management of natural resources and for the improvement of agricultural production systems has been increasing. The establishment of the Central American Commission on Environment and Development (CCAD) reveals the importance of this issue for public policy. Moreover, the large number of local initiatives and ongoing rural development programs from governmental and non-governmental organizations offer considerable experience on practical alternatives for sustainable resource management.

The contributions in this reader offer a comprehensive review of the current prospects for sustainable agriculture as assessed from state policies and local action. For this, trade-offs between macro-economic policies and environmental objectives are determined, and the socio-economic conditions for the establishment of sustainable production systems in different eco-regional settings (hillsides, humid tropics, frontier areas) are identified. Special attention is given to the various institutional arrangements that are used to guarantee the conservation of nature areas. Finally, the policy instruments available to improve property rights, and the management regulations and financial mechanisms available to enhance sustainable resource use are discussed. The salient conclusion that emerges from the contributions indicates that in Central America there is still a long way to go before effective coordination between the state and private initiatives may permit the process of resource depletion to be effectively controlled.

1

2 SUMMARY

The macro-economic conditions for sustainable agriculture are discussed in Part I. In terms of policy objectives, the sustainability of the resource base in Central America had a low priority until recently. The picture of contemporary development that emerges is one of society and state disregard for sustaining the resource base, as reflected in (i) the breakdown of traditional resource management systems that were ecologically sustainable, and (ii) the unchecked clearance of forest land and inappropriate land use practices and technologies.

But although the contemporary state of degradation of resources can be traced back to certain macro-economic policies – related to inadequately developed legal frameworks, economic incentives, institutional failures and cultural attitudes – it is far more difficult to generalize about the impact present macro-economic reforms are having on the environment. For one thing, production is now much more sensitive to movements in factor and commodity prices, and policies presently encounter a much wider array of potential production decisions. For another, whereas governments now promote environmental protection, at the same time they are implementing policies that result in resource degradation, or that weaken the institutions responsible for protecting and rehabilitating of the environment.

Conditions for the introduction of environmental protection schemes and sustainable land use systems and practices include not only tracing how policies at the macro-level affect resource use, but also not ignoring the contradictory effects of government policies on the environment. Sometimes the effects of macro-policies are straightforward; as for example in the case of the impact of road construction in the agrarian frontier on deforestation. In other cases it is more complicated to trace the effects on resource use. Contradictory effects of policies develop when the government does not address structural inequities in the distribution of income and wealth, or even allows structural adjustment policies to exacerbate them, thereby adding extra pressure on the natural resource base by increasing poverty. Such contradictions also develop when policies for economic stabilization impose constraints on environmental protection agencies. Tracing the effects of macro-economic policies on resource use and redressing the contradictions are particularly important conditions for a more participatory approach to resource protection that takes into account the rights, needs, cultural practices and livelihood concerns of local resource users.

The theme of Part II is sustainable production systems. It is emphasized that the variety of production systems is much wider than could be explained by a dualist view. Traditional systems, whether hillside farming, semi-arid agriculture or farming in the humid tropics, are continuously involved in processes of change. Producers have responded in many ways to the challenges of the local ecology, climatic factors and soil conditions, in the context of social and economic change, and they have created different constellations of technology and agricultural practices. An empirical description of a marginal hillside area in Honduras is presented showing the wide variety of production systems that have developed in response to biophysical production constraints and social, economic and technological changes.

Sustainability is viewed more broadly than in ecological terms; it is – including in some of the papers in this part – defined as a process of change to build human capacity and to create a resource base to meet human needs. Food insecurity of Honduran petty producers and semi-proletarian farmers is used as a case to present points of vulnerability created by exchange relations. The exchanges – land, credit and output – can serve both to undermine or to enhance farmers' capacities to continue in production, as can interventions by the state and other institutions in a context of unequal social relations.

In the humid tropics of Nicaragua the limits of the agrarian frontier are about to be reached. Land is becoming relatively scarce both for the individual colonist trapped in a low-level equilibrium, and collectively. Accordingly, extractive production systems will have to be replaced by more sustainable ones that reverse the falling productivity and restore degraded land. The main question policy must answer is how to promote the transition to such alternative land use systems and practices. In the older colonization areas livestock development on smaller farms and improved crop yields – both attained using available and rather simple agricultural practices – can play an important role, as can the improvement of market organization. In the recent agrarian frontier options to prevent land degradation and the fall in productivity still exist; these will make extensive cattle ranching less attractive.

The most important change in land use in Central America in the last 40 years has been the widespread conversion of forest to pasture, the development of (extensive) ranching. However, deforestation declined significantly in the late 1980s. Unfavourable market conditions for meat and dairy products, reduced access to credit for livestock and higher interest rates were limiting factors for the livestock sector. But they probably had a larger impact on the investment ranchers and the large cattle owners in the traditional livestock regions outside the agrarian frontier than they had on pasture expansion in the agrarian frontier areas.

To reduce deforestation in the agrarian frontier, road construction and improvement will have to be discouraged and the public lands which cannot be allowed to pass into private hands will have to be defined – and these decisions will have to be enforced. It will also be necessary to eliminate the incentives to clear forest in order to claim land or to improve tenure security. The protected areas must be protected, while at the same time establishing the best possible relations with neighbouring communities, in particular with the indigenous communities.

Part III deals with natural resource reserves and protected areas. In most of the natural resource reserves in Central America there is a steady inflow of colonists driven by poverty and landlessness – a consequence of national and international policies – that exacerbates competition for land and resources between indigenous communities and *mestizo* inhabitants. Moreover this colonization threatens attempts by governments to protect reserves and their core from deforestation. To understand the constraints to sustainable development in natural resource reserves there is necessary to examine conflicts about land and resources between indigenous communities and the *mestizo* population; to explore differences between indigenous and *mestizo* environmental perceptions and land use practices; and to consider tensions between peasant and indigenous livelihoods and conservation. As discussed in one of the papers in this Part, in Nicaragua's Bosawas Reserve the land claims situation is complex. The challenge facing the policy makers is to safeguard indigenous land rights while gaining the co-operation of the *mestizo* inhabitants; and to work with both groups towards the sustainable management of the reserve.

The strategy for sustainable management of natural resource reserves, and even of the core, the protected area, cannot be based solely on 'fences and fines' type of measures. The drawback of this approach is that protection by exclusion of people reduces a forest's economic use value. If they do not share in forest management that provides a cash income, most local communities do not perceive incentives to conserve the forest; clearing the forest to use the land for agriculture seems the only benefit to be gained. There has been some outstanding work by ngo's particularly in the development of sustainable farming systems in buffer zones. But considering that the land is often really of forest vocation, it is not enough that people are made better off by developing sustainable farming systems; the benefit must be in a form that induces them to conserve the forest. There is a need for policies and incentives that promote positive attitudes to natural resource conservation and management. Development and conservation can be linked through eco-tourism and community based natural forest management that provides a cash income.

In Part IV agrarian policies for sustainable land use are discussed. The crisis of peasant agriculture in Central America is mainly one of institutions in the countryside. Peasant production systems were expected to benefit from liberalization and elimination of price distortions under structural adjustment programmes. But although the previously existing economic institutions have been disrupted, no new ones have yet developed. In Nicaragua the profound transformation of the agrarian structure has undermined the former well-articulated socio-economic structure. Economic crisis required economic reforms in all the countries of Central America. The programmes of economic reform (economic stabilization, structural adjustment) also disrupted existing institutions, partly because reforms of commoditized markets often have a major impact on non-commoditized exchange relations based on personalized networks. Liberalization does not automatically imply the disappearance of market failures. For that to occur, new institutions have to develop that replace the economic functions of the old structures, articulating viable national-level structures with appropriate local organizations.

One of the areas discussed in relation to this is the definition of rules for resource management in agrarian production cooperatives, APCs, with a case study from Honduras. The management of common property resources has for long been considered as a device for inefficient allocation and for resource depletion, as problems of free riding and monitoring tend to promote non-cooperative behaviour. Within the property rights paradigm external enforcement is required to overcome these problems, and thus privatization is promoted as a device to improve resource allocation according to opportunity costs. This approach is considered insufficient, as cooperation emerged precisely as a substitute for missing markets. It is possible to improve technical and allocative efficiency within APCs if incentives are adequately addressed. Contract-choice theory offers new perspectives for the appraisal of feasible *internal* incentives and operational rules that enhance efficient resource allocation and sustainable land use within a cooperative context.

The property rights literature suggests that unambiguous property rights, land titles, are the key factor for sustainable use of land. This is the subject of another paper in this Part which reports the experiences from a case study in Honduras. The study shows that land titles are appreciated by farmers, but that other factors such as credit and agricultural extension are at least as important. Isolated land titling projects that are not accompanied by other development activities are unlikely to achieve their objectives. That land titling has only a partial influence on land use systems is a consequence of the gradual change in the property situation; land titling as a national system is replacing local and semi-official arrangements which already assured certain levels of tenure security. Farmers, moreover, assess conservation measures in economic terms and will only use conservation techniques if they prove cost effective.

A second part in this Part deals with institution building for (equitable) rural development in Nicaragua, emphasizing the strategic role of non-conventional finance. After more than a decade of profound structural reforms, Nicaragua is still searching for a new institutional articulation of its rural economy. One of the main problems is clearly the relative incompatibility between the vertical-clientelistic institutional heritance and the objective equality of the agrarian structure of the country. A transition toward more horizontal governance structures would clearly be preferable from a variety of perspectives. However, operational and 'power-compatible' strategies need to be found and put into practice in order to reach such a transition. The creation of local institutional structures for finance in the midst of an extreme liquidity crisis could provide a means to promote more civic-horizontal rural institutions, starting with the institutionality for finance. This will require a broad strategic perspective on non-conventional finance that transcends the mechanics of making a non-conventional financial enterprise work in the short run.

The main advantage of non-conventional credit programmes is their great flexibility that enables them to adapt their schemes to a target group. These programmes have more incentives than the development banks to introduce institutional innovations in the domain of the screening of clients and recovery strategies. If they succeed in extending access to credit for poor households and in improving their financial performance donor organizations reward them with concessional credit or donations. The paper on this topic compares experiences of performance of a number of rural lending projects in Central America with the former experiences with agricultural development banks.

Part I

Macroeconomic Conditions for Sustainable Agriculture

2 Deforestation in Central America: Historical and Contemporary Dynamics[1]
Peter Utting

1 INTRODUCTION

When reading about deforestation in Central America one is usually presented with a list of 'causes' which is likely to include the colonization of agrarian frontier regions by land-seeking peasants and commercial farmers, slash and burn agriculture, the so-called 'hamburger connection' involving the rapid conversion of forest land to pasture, the expansion of cash crops such as coffee and bananas, logging activities, fuelwood gathering, and additional pressures on forest land exerted through population growth and urbanization. Many of these 'causes' of deforestation, however, are really features of a particular style of development that began to take shape during the colonial era, became more pronounced during the late 1800s and 'took off' during the latter half of this century.

The purpose of this paper is to set the phenomenon of deforestation in the context of the development model that has evolved in Central America, particularly during the past century. The paper is divided into five main sections.

Section 2 describes the extent of deforestation in the region and briefly indicates why deforestation has become a major development problem in recent decades.

Section 3 examines the relationship between deforestation and agro-export development, focusing in particular on the way in which three processes of change, associated with the so-called agro-export model, have underpinned deforestation.

Section 4 examines the process whereby certain local resource management systems, which to some extent used the region's natural resources in a more sustainable way, have been weakened or have

9

collapsed altogether, to be replaced by systems that degrade the environment and which imply increasing material and cultural insecurity for certain social groups.

Section 5 identifies a number of contemporary developments that are altering the dynamics of deforestation in the 1980s and 1990s.

2   THE SCALE OF THE PROBLEM

Deforestation as a numbers problem

Throughout the 1960s and 1970s, Central America experienced one of the highest rates of deforestation in the world. Since 1960, the extent of forest cover in the region has declined from approximately 60 per cent to a third of the total land area.

Estimates of the extent of forest cover and rates of deforestation in the region vary considerably. Figures on annual deforestation in some countries differ by as much as 100 per cent. New sources of information based on satellite imagery are leading to some significant revisions in estimates. Data presented in Table 2.1, derived mainly from official sources, indicate that by 1990 the region was losing about a third of a million hectares of forest or woodland each year or nearly 2 per cent. This data generally refers to both closed and degraded forest cover.

There are indications that the rate of deforestation has declined somewhat during the 1990s. One recent study of land use in Central America has put the loss of forest area at around 300 000 ha. per annum (Kaimowitz, 1995). Some sources, however, still cite figures in excess of 400 000 ha. (see Pasos *et al.*, 1994, p. 3).

But three developments, in particular, suggest that the overall figure may have declined. Firstly, new data for the Petén, where the bulk of Guatemala's forest resources are concentrated, suggest that deforestation in that country is likely to be in the order of 50–60 000 ha. a year (Kaimowitz, 1995), as opposed to figures of 90 000 or more which are often cited. The FAO has recently revised its estimate of the extent of forest cover in Guatemala upwards by nearly 30 per cent from 4.5 million ha. in 1983 to 5.8 million in 1993 (FAO, 1995). Secondly, for reasons identified in Section 5 of this paper, there has been a significant decline in the rate of deforestation in Costa Rica since the late 1980s. Some commentators suggest that the current annual rate of deforestation may have fallen below 10 000 ha., compared with 50 000 ha. in the late 1970s (Kaimowitz, 1995 citing Lutz *et al.*, 1993). Thirdly, there are indications

Table 2.1 Central America: national estimates of forest cover and annual
deforestation, 1990

country	area		% of total land area	annual deforestation (ha.)	
Panama	3203	(1)	42	34 000	(7)
Costa Rica	1476	(2)	33	50 000	(8)
Nicaragua	4140	(3)	30	70 000	(9)
Honduras	4731	(4)	42	80 000	(10)
El Salvador	250	(5)	12	14 000	(11)
Guatemala	3762	(6)	35	90 000	(12)
Total	17 502		36	338 000	

1. Estimate for 1990 based on 1987 figure of 3 305 300 ha. cited in INRENARE, 1991:13.
2. Estimate for 1990 based on 1989 figure of 1 475 940 ha. cited in MIRENEM, 1990.
3. Ministerio de Agricultura y Ganaderia (MAG), 1990.
4. Estimate for 1990 based on COHDEFOR (SECPLAN *et al.*, 1990)
5. Mansur, E., 1990.
6. Plan de Accion Forestal para Guatemala, 1990.
7. INRENARE, 1991:13;
8. Bonilla, 1988:72; MIRENEMA, 1990:4;
9. MAG,1990;
10. SECPLAN *et al.*, 1990;
11. Mansur, E., 1990;
12. Bradley, T., *et al.*, 1990a.

that considerable areas of agricultural land are reverting to scrub, woodlands or secondary forests of one sort or another as pastures are abandoned for both economic and ecological reasons (Kaimowitz, 1995; Pasos *et al.*, 1994). Kaimowitz indicates that one-sixth of Nicaragua's land area, or 2 million hectares, consists of 'scrub forest' or 'forest fallow', and suggests that in Costa Rica the area under secondary forest may have nearly doubled from 229 000 to 425 000 ha. between 1984 and the early 1990s. It cannot be assumed, however, that rates of deforestation have declined in all countries. Indeed, it is generally assumed that deforestation in Nicaragua is once again increasing following the ending of the war which had impeded logging activities and the conversion of forest to agricultural land.

Deforestation as a development problem

The environmental impact of deforestation varies considerably according to the very different climatic and soil regimes which exist in the region. The Pacific slope and coastal areas are characterized by a clearly defined wet and dry period, while the Atlantic lowlands experience rainfall during most of the year. Large areas of the Pacific-coastal and hilly interior regions of Guatemala, El Salvador, Honduras, Nicaragua and Costa Rica have relatively fertile volcanic soils which are prone to wind and water erosion once forest cover is removed. Soils found in much of the Caribbean-coastal strip, as well as in northern Guatemala, often have a very thin topsoil layer and are prone to leaching, particularly when the removal of tree cover accentuates the force with which rain impacts the ground (Leonard, 1987, pp. 13–14; Brüggemann and Salas, 1992). Throughout the region, deforestation has caused major problems of soil erosion, and flooding due to increased run-off and sedimentation of water sources. Significant changes in micro-climate have also occurred, some-times producing or intensifying drought conditions.

The social and economic ramifications of deforestation are equally dramatic. Deforestation threatens the livelihoods and lifestyles of much of the region's population in numerous ways. Fuelwood, for example, is used by nearly three-quarters of the region's households. The rise in fuelwood prices and the increasing difficulties of gathering fuelwood have had damaging effects on the household budgets and workloads of many families. Accessing sufficient quantities of water both for agriculture and household needs has also become a major problem in both rural and urban areas. The destruction of mangroves threatens the viability of some fishing communities. The decline in forest areas and wood supplies has exacerbated social tensions and there are increasing violent conflicts centred on the use of land and forest resources. As access to forest resources becomes more difficult many people continue to exploit forest resources through clandestine means. This growing phenomenon of illegality strains the social fabric even further and undermines the ability of the state and external agencies to intervene in the development process.

3   DEFORESTATION AND AGRO-EXPORT DEVELOPMENT

Throughout the colonial period, and even during the early decades of the republican era in the first half of the nineteenth century, the agrarian frontier advanced relatively slowly. Spanish control and colonization of much of the region's more inhospitable areas along the Caribbean coast, with dense forests and high rainfall, was limited. Indeed attacks by pirates, Sambos and miskito indians throughout the 18th century actually caused the agrarian frontier to be pushed back in some regions such as Nicaragua (Pasos *et al.*, 1994, p. 21).

Agro-export growth

The contemporary phenomenon of rapid deforestation is a feature of a particular style of development based on the production of products such as bananas, coffee, cotton and beef destined largely for the international market. Certain farming systems, notably ranching, devoured large areas of forest since they were based on a pattern of 'extensive accumulation' (Baumeister, 1991) where both land to labour and land to input ratios were extremely high.

Agro-export growth accelerated during the 1950s and 1960s and was accompanied by the rapid conversion of forest to crop land and pasture. Economic growth and modernization was also accompanied by an intense process of social marginalization which forced many peasant families to migrate to agrarian frontier regions in search of land and food.

Underpinning this model were powerful political forces that ensured that national states actively promoted agro-export expansion (Carriere, 1991). Government policies facilitated the assault on the forest through the provision of incentives, services and infrastructure for export production, as well as legislation conducive to the rapid colonization and logging of forest areas. National states also attempted to maintain social stability and defuse unrest, sometimes through authoritarian means, but also by using forest land as a safety valve.

Throughout this century, the primary pressures on the natural resource base have been intimately related to certain processes of economic and social change that characterize the so-called agro-export model. Three processes are particularly relevant: *market integration, modernization* and *marginalization*.

Market integration

Market integration is relevant in two major respects. Firstly, national economies were drawn into the world market for primary goods, notably coffee, cotton, sugar, bananas and beef. In several countries, this process accelerated during the latter half of the last century when world demand for Central American coffee rose sharply. Given the association of coffee with shade trees, the scale of environmental deterioration related to coffee expansion was not as intense as that which occurred during the second half of this century when the economies of the region experienced rapid growth in several agro-export product sectors.

Particularly dramatic was the environmental impact of the cattle boom. The clearance of forest areas increased sharply in Honduras, Nicaragua and Costa Rica during the 1960s as demand for Central American beef increased in the United States (Nations and Komer, 1987). From the mid-1950s to the mid-1970s, the area under pasture in the region increased from 3.9 million to 9.4 million hectares, or nearly one fifth of the total land area (Heckadon, 1984; Williams, 1986).

In large areas, the expansion of pasture area did not mean, however, the destruction of primary or secondary forest but the reduction of fallow land which was crucial for the sustainability of farming systems characterized by shifting agriculture. As such, the cattle boom should be associated not simply with extensive deforestation but also the breakdown of traditional resource management systems.

But market integration also operated at another level. Peasant producers were drawn increasingly into national markets. The intensification of commodity relations meant that rural families not only had to produce much of the food they required but also a marketable surplus in order to obtain the income necessary to purchase production inputs and consumer goods and services. This led to an intensification of agricultural production which in many areas of Central America broke the fragile ecological equilibrium that characterized the traditional slash and burn system whereby land had to be left fallow for many years before it could be cultivated for relatively short periods. It also accelerated the conversion of forest areas to crop and pasture land.

Modernization

Modernization implied a certain vision of progress which saw the forest as an obstacle to development. Until fairly recently, government policies and laws openly encouraged the colonization of rainforest areas. In countries such Nicaragua, Honduras and Guatemala, governments implemented large planned colonization schemes during the 1970s or 1980s. In Costa Rica and Panama, as in most other countries, government polices actively supported 'spontaneous' colonization. Laws often required colonists to clear the forest in order to establish the right of possession.

But many who colonized rainforest areas did not acquire secure title to the land. The itinerant nature of farming systems in some frontier regions, the low value of land, as well as complicated, costly and time-consuming legal and administrative procedures meant that few bothered to obtain legal title (Augelli, 1987; CEDARENA, 1990).

Insecurity of tenure has important implications for deforestation; if possession could not be proved legally, it had to be demonstrated visually, by clearing forest.

The process of acculturation which is a feature of modernization has considerably weakened traditional beliefs regarding the sanctity of the natural elements. It has also weakened the extremely rich knowledge base which underpinned the sustainable use of forest resources and enabled indigenous groups to use numerous forest products for basic needs provisioning.

The culture of modernization encouraged the introduction of new technologies that were often inappropriate for the type of ecological conditions prevailing in tropical forest areas and implied greater risk for the producer. The introduction of modern technologies sometimes had disastrous environmental and socio-economic consequences for certain farming systems. Such was the case, for example, in several coffee producing areas of Costa Rica and Nicaragua when, during the late 1970s and 1980s, shade trees were eliminated from many coffee farms to make way for coffee varieties that demanded large quantities of inputs.

Modernization meant, for certain groups and geographical areas, improved health care facilities and reduced levels of infant mortality which in many rural areas meant larger families to feed, clothe and care for and, hence, increased demands on the land and agricultural production.

Modernization also implied the development of economic infrastructure such as roads, railways and hydroelectric power schemes. The expansion of the public road network greatly facilitated the colonization of forest areas. Particularly important during the second half of this century was the construction of the Inter-American highway that aimed to link North and South America. In Costa Rica it has been estimated that the rate of deforestation on the Pacific side of the country increased five-fold following the construction of the highway (Silliman, 1981, p. 65). The generous concessions granted to the banana companies carried with them the commitment to construct the railroads. As such deforestation was partly associated with the demand for railroad sleepers.

Marginalization

Marginalization refers to highly skewed patterns of resource distribution that have characterized the agro-export model and left much of the region's population living in extreme poverty. It also refers to processes of 'disempowerment' that result in the loss of rights and reduced control for large social groups over resources and decision-making processes that are related with resource management (Vivian, 1991).

The improved opportunities for commercial farming which accompanied the insertion of the Central American economies in world commodity markets prompted processes of land concentration and the displacement of peasant producers from the land. Interests (both national and foreign) involved in the production, processing, and trading of agro-export products, as well as the financing of such activities, came to exert a dominant influence over national states. Government policies and development programmes generally favoured such groups and largely ignored or discriminated against peasant producers and indigenous groups.

Much of the rural population has insufficient access to arable land, as well as credit and other resources needed to sustain a farming operation. Peasants would often migrate to agrarian frontier areas. Some would settle on forest land belonging to ranchers, agree to clear it, farm it for one or two seasons and then leave it seeded for pasture before moving on to clear another plot. Others would settle on unclaimed land, produce food crops for a short number of years and then be forced to abandon the plot due to declining fertility and weed growth. Instead of leaving the land fallow, however, they would seed it for pasture, sell the 'improvements' to ranchers, and then move on.

Given the extremely low labour requirements of the cattle sector and the fact that the expansion of pasture land in many areas displaced more labour-intensive forms of agricultural production, the possibilities of obtaining additional income from wage employment were often limited (Howard, 1987).

For many peasant families, subsistence provisioning became increasingly difficult given the shortening of fallow periods and the need to farm poorer quality land, or due to the difficulty of finding employment. Migration to urban areas or the agrarian frontier became the logical option for hundreds of thousands of peasant families throughout the region.

Pressure on forest areas and the breakdown of traditional resource management systems are intimately linked with processes of land concentration and *minifundismo*.[2] So-called 'liberal' states that emerged during the late 1800s and early 1900s fostered the growth of coffee production and introduced laws that tended to accelerate the disintegration of communal holdings. The profit opportunities associated with the cotton, beef and sugar booms of the 1950s, 60s and 70s, further intensified processes of land concentration and landlessness. By 1970, approximately half of all rural families were either landless or farmed sub-subsistence plots of less than one hectare (Weeks, 1985).

4 THE BREAKDOWN OF TRADITIONAL RESOURCE MANAGEMENT SYSTEMS

Historically, sustainable farming or resource management systems have been practised throughout the Central American isthmus both by indian groups and *ladino* peasant farmers. Such systems combined agricultural and forestry systems that were suitable from an ecological point of view and provided a degree of 'social security' for local farming populations. Many such systems, however, have long since collapsed. Others are currently breaking down when confronted with the effects of market integration, acculturation, population growth, economic crisis and government policies related to human settlement, land tenure and economic development.

This section analyses three specific resource management systems involving indigenous populations and *ladino* peasants in three different countries.

Deforestation, encroachment and traditional forest dwellers

Pre-Colombian Central America was home to numerous indian groups who engaged in sustainable forms of subsistence provisioning. Pressures on the natural resource base were minimized through a combination of conditions related to low levels of surplus extraction (due primarily to the absence of markets), low person to land ratios, as well as cultural perceptions and religious beliefs that led people to respect nature's elements.

Agricultural practices based on long crop-fallow rotations ensured the regeneration of forest areas cleared for cultivation and the recuperation of soil fertility. Shifting agriculture was generally combined with hunting, gathering and fishing. Since the arrival of the Spaniards in the sixteenth century, these systems have been under threat. Many indian groups were forced to migrate from coastal or lowland areas to more inaccessible highland regions. Despite the effects of colonization and modernization, several indigenous groups still practise sustainable resource management systems combining agriculture, agriculture and wildlife management which produce a multiplicity of products necessary for subsistence provisioning and commerce (Houseal *et al.*, 1985).

The stresses imposed upon such systems have intensified dramatically during the latter half of this century due primarily to encroachment by outside forces. When analyzing the impact of encroachment on indigenous groups, two scenarios stand out: one involving settlement by ranchers and *ladino* or white peasant farmers, and another involving the operations of large 'extractive' enterprises usually engaged in logging or mining activities. Encroachment has reduced the areas available for sustainable shifting agriculture based on long crop-fallow rotations, led to displacement of indian families to more marginal areas, and transformed social relations of production and indigenous culture.

Very often indians were displaced to more marginal and inaccessible areas characterized by higher rainfall, steeper slopes and poorer soils (Houseal *et al.*, 1985, p. 10). Not only are many indian families finding it increasingly difficult to practise sustainable resource management as land becomes increasingly scarce but indigenous communities are rapidly losing the knowledge base associated with such practices. This process of encroachment has forced much of the indian population to abandon traditional agricultural or subsistence provisioning practices based on long crop-fallow rotations and hunting and gathering.

The ladino 'roza' system

Throughout many centuries, peasant agriculture in Central America has been based on the slash and burn or *roza* system. As Heckadon points out, such a system can yield important social, economic and ecological benefits under conditions where the person to land ratio and the cash and consumer demands of the peasant household are low (Heckadon, 1982). It not only provided peasant families with their basic food requirements but also minimized risk given that few or no costly modern inputs were required and indebtedness was restricted. By burning the dense covering vegetation of forest areas the peasant transformed this biomass into nutrient rich ashes which fertilized crops. The *roza* system, however, depended on the possibility of leaving land fallow for periods of 10, 20 or more years given the impossibility of continuing production on the same plot under conditions of declining fertility and prolific weed growth. The natural regeneration of forest served to increase the fertility of soils, improve their structure, protect them from erosion and reduce the incidence of weeds (CIERA, 1984, pp. 83–7; Heckadon, 1984, p. 218).

This system, however, is extremely fragile and liable to break down as pressure on the land and the rate of deforestation increase. When this occurs, the impact on agricultural production and productivity, employment, livelihood and social relations can be dramatic. These relationships have been analyzed in some depth by Heckadon in his studies of colonization processes in Panama (Heckadon and Mckay, 1982; Heckadon, 1984; Heckadon and Espinoza, 1985).

Deforestation and communal forest protection[3]

Over several millenia, many indigenous communities in Central America evolved and adapted communal institutions for regulating the use of forest resources. Many such systems have long since been displaced or eroded by market forces or state legal codes.

A system of communal forest protection that proved to be particularly effective, even in the modern era, operated in the department of Totonicapán in the western highlands of Guatemala. In this region many families own extremely small plots of land, often less than one-third of a hectare. Population density is of the order of 260 inhabitants per square kilometre. Despite considerable pressure on the land, however, much of Totonicapán did not experience rapid deforestation until fairly recently. This was largely due to the effectiveness of a variety of communal regulatory mechanisms employed by indian communities in the department (Veblen, 1978).

Faced with a decline in subsistence provisioning through direct agricultural production and restricted employment and income-earning opportunities in artisanry, trade and other activities, many families have turned to the exploitation of forest products as a means of finding supplementary income or reducing household expenditures on wood products. In other words, many families turn to the forest as part of a 'coping strategy'. Such activities include primarily fuelwood gathering, wood-sawing, bark-stripping, pine resin extraction, and the cutting of resinous pine splinters used for starting cooking fires. Trends in relative prices have also contributed to the increased exploitation of forest resources.

Forest protection mechanisms have weakened as a result of changes in land tenure. Through time, communal forest areas have gradually been privatized or come under the control of municipal authorities. Rates of deforestation are often higher in private and municipal forests and when the status of communal lands is threatened and tenure disputes arise, the system of communal forest protection tends to break down.

The weakening of the communal protection mechanisms is partly accounted for by the changing nature of economic and subsistence activities associated with the exploitation of forest products. Many activities have assumed a clandestine character which renders detection increasingly difficult. To avoid detection people often work at night.

5   THE CONTEMPORARY DYNAMICS OF DEFORESTATION

Crisis and restructuring

Throughout much of Central America, the decade of the 1980s was syn-onymous with economic crisis and war. Economic and social life in the region was also affected by structural adjustment of one type or another. This has usually involved restructuring along 'neo-liberal' lines – involving conventional economic stabilization and adjustment policies. But in Nicaragua and El Salvador restructuring also assumed a redistributive cha-racter, involving agrarian reform. All these phenomena – economic crisis, war and restructuring – have affected the dynamics of deforestation.

Stagnant or falling international prices for sugar, cotton and beef and restrictions on Central American imports into the United States put a brake on the expansion of traditional agro-export agriculture. A positive deve-lopment from the ecological point of view has been the fact that the rate of expansion of pasture area declined considerably in all countries (Baumeister, 1991; Kaimowitz, 1995), while the total area under pasture in Central America actually appears to have fallen between 1983 and 1991 due to a significant drop in pasture area in Nicaragua (Kaimowitz, 1995). These trends are partly explained by the fact that real prices for beef on the international market halved between 1970 and 1990 (Kaimowitz, 1995, citing Trejos, 1992).

While the agro-export recession no doubt served to constrain the process of 'extensive accumulation' noted earlier, it may well have obliged thousands of families to intensify the exploitation of forest resources as a form of coping strategy. Relevant in this context is the fact that recession in agriculture has constrained the process of 'proletarianiza-tion' of the rural labour force (Baumeister, 1991). In several countries, there has been a marked decrease in the level of demand for harvest labour in the agro-export sector. In such contexts, it can be expected that many people will look to forest resources or the clearance of forest land as a means of acquiring income and subsistence goods.

The military conflicts of the 1980s in Nicaragua, El Salvador and Guatemala were to have important effects on deforestation. The precise nature of these impacts, however, has varied by country. In El Salvador and Guatemala, military operations caused major degradation of some forest areas. Armies in both countries operated a scorched earth policy in several areas considered guerrilla strongholds. In Nicaragua, fighting in the interior and Atlantic coast regions acted as a brake on the expansion of the agrarian frontier and seriously damaged the cattle industry. In all three countries hundreds of thousands of people were displaced from their homes. Deforestation usually intensified in the areas where refugees or displaced persons settled. The resettlement process of the 1990s that followed the cessation of hostilities has had a similar effect. In El Salvador and Nicaragua significant agrarian reform programmes were implemented. In certain areas it can be assumed that providing improved access to land or security of tenure served to reduce pressures on the forest associated with colonization and shifting agriculture. But in both Nicaragua and El Salvador, land redistribution also accelerated processes of forest clearance in many areas subject to reform.

But restructuring in Central America has generally been of a very different kind, following the neo-liberal route of economic stabilization and adjustment. The changing patterns of resource allocation associated with such policies and programmes have important implications for deforestation. Partly as a result of these policies, so-called 'non-traditional' agro-export sectors (flowers, spices, vegetables, sesame, cocoa, etc.) have expanded. To the extent that this pattern of development promotes a more intensive, as opposed to extensive, model of accumulation then it is possible to theorize that structural adjustment may diminish the pressure on forest resources. In practice, however, there are many ecological question marks associated with the patterns of agricultural production typically associated with structural adjustment (Reed, 1992).

Economic stabilization programmes, intent of cutting fiscal deficits, have imposed restrictions on government expenditures. In some cases this has severely limited the budgets of environmental agencies and their capacity to implement environmental programmes and enforce regulations governing natural resource use. Many forest protection projects and extension services experience major difficulties due to a shortage of working capital or supplies.

Conservation and 'eco-capitalism'

The Central American development model in general, and the dynamics of deforestation in particular, have been changing in recent years as a result of two other developments that would seem to augur well for the environment. One relates to the increasing attention on the part of government, development agencies and NGOs to conservation and environmental rehabilitation. The other, concerns the rise 'eco-capitalism' with sectors of the business community becoming actively involved in the production and trade of environmental goods and services. Concomitant with these trends, and indeed reinforcing them, has been a shift in the balance of social and political forces which has seen an increase in the power and influence of interests which favour certain forms of environmental protection.

The upsurge in environmental awareness and in the number of environmental NGOs, coupled with the increase in environment-linked aid, has prompted some significant changes in environmental policy and a marked increase in conservation programmes and projects. A number of important initiatives involving forest protection and tree planting have been taken in most countries.

During the past decade there has been a flurry of activity on the part of several Central American governments, international agencies and national NGOs aimed at designing and implementing policies, programmes and projects to conserve or rehabilitate the natural resource base. The number of wildland areas officially declared as *national parks and reserves* increased. While accurate data are difficult to come by, it would appear that over 8 million hectares, or one-sixth of the region's total land area, have, or were about to receive in the early 1990s, protected area status.

A number of governments have introduced credit and fiscal incentive schemes to promote *reforestation*. The 1980s saw a sharp increase in the number of *agroforestry and social forestry projects*. Moreover, considerable resources have been channelled towards research and experimentation on agroforestry systems, notably through the work of the Tropical Agricultural Research and Training Centre (CATIE). In countries such as Nicaragua and Honduras, so-called 'farmer-to-farmer' training and extension programmes have also been effective in promoting sustainable natural resource management and protection among peasant farmers and cooperatives in forest areas (Pasos *et al.*, 1994).

Since the late 1980s, there has been more attention to the question of how to promote 'sustainable logging' although achievements on the ground have lagged far behind the reformist discourse. Increasingly, however, the crucial question of how to integrate local communities in forest management schemes is being posed and a number of projects have encouraged this approach.

The process of *environmental planning* has been assisted by the elaboration of studies and plans in most countries consisting, for example, of Tropical Forestry Action Plans, Natural Resource Policy Inventories and National Environmental Action Plans. Regional co-operation to protect and rehabilitate the environment has increased following the formation of the Central American Commission on Environment and Development (CCAD) in 1989 (Arias and Nations, 1992). These developments have served to intensify government and NGO efforts in this field and attract foreign aid to support conservation projects and programmes. Regional co-operation, involving not only the Central American countries but also the United States, reached new heights in 1994 with the proposal to forge an 'Alliance for Sustainable Development'.

There are indications that certain sectors of private enterprise are 'accommodating' to environmentalism and to new market opportunities or constraints associated with environmental goods and services. In several countries private business interests are engaging in, or are supporting, reforestation, forest plantations, protected area schemes, wood certification programmes, eco-tourism, trade in organically grown products, and so forth.

It is possible to identify at least six developments associated with eco-capitalism that are affecting to varying degrees the dynamics of deforestation. As the following examples suggest, the development of 'eco-capitalism' is far more advanced in Costa Rica but is likely to become more relevant in other countries as well.

- Since the late 1980s, ranchers and lumber companies in several countries, notably Costa Rica, have begun to participate in fiscal incentive schemes to reforest pasture or woodland. The 1986 Costa Rican Forestry Law provided greater incentives for *reforestation* on private land by introducing transferable bonds for reforestation projects that can be used to pay taxes or sold at slightly less than face value (Bradley *et al.*, 1990b).
- Several companies have taken advantage of certain government incentives and new market opportunities to invest in forest *plantations*. The area under forest plantations in Costa Rica has increased rapidly in recent years, from less than 30 000 hectares in the late 1980s to approximately 90 000 in 1994 (Kaimowitz and Segura, 1995).
- There has been considerable growth in business interests associated with tourism and, in particular so-called 'eco-tourism', centred on national parks. In the space of less than a decade, tourism in Costa Rica moved from being a relatively marginal foreign exchange earner to become the largest. As a result of this development there now exists a powerful sector of business with a direct interest in supporting the national policy to consolidate the country's system of protected areas.
- So-called 'eco-labelling' appears to be on the increase. In response to both the threat of legislation to ban logging outside of forest plantations and tighter restrictions in Europe on the importation of wood harvested by unsustainable logging practices, the Costa Rican lumber

association which represents the country's logging companies has proposed a sector-wide wood certification scheme.

- There are signs that Northern transnational corporations in the pharmaceutical and agrochemicals industries are taking greater interest in the search for genetic resources in the region's tropical rainforest areas. The increasing benefits that can be derived from 'bio-prospecting' have led some transnational corporations to take a more active role in supporting rainforest protection and protected area schemes. Underpinning this development is the increasing value of bio-diversity which has resulted from the rapid disappearance of rich sources of germplasm, advances in bio-technology which enable scientists to isolate genes and transfer them to other organisms, and the consolidation of an international system of intellectual property rights (Cabrera Medaglia, 1995). Costa Rica is the site of a much publicized partnership involving the national government, the conservation NGO INBio and the US pharmaceutical company Merck.

- Support for reforestation, agroforestry and soil conservation projects is also coming from northern energy corporations. One US company in Guatemala, for example, has a 'carbon offset programme' which provides funding for environmental protection projects as a means of offsetting the company's carbon emissions in the United States. Under terms agreed in the Framework Convention on Climate Change energy companies can obtain credits on carbon emissions as a result of their investment in forest protection and treeplanting schemes that sequester carbon from the atmosphere through photosynthesis and tree growth.

Initiatives such as these are not only changing the natural landscape through reforestation, forest plantations, agroforestry and protected areas, they are also providing valuable resources to government agencies and NGOs involved in environmental protection. And politically, certain corporate interests are adding their considerable weight to the environmental 'movement' or lobby and influencing government policy.

But there are many question marks surrounding these initiatives and their ability to protect or rehabilitate the region's forest resources and promote sustainable development. The most obvious concerns their scope. The record of modern-day conservation in the region reveals numerous failed or unsustainable projects, so-called 'paper parks' and environmental protection plans and strategies that simply gather dust. But of equal concern are a series of effects that may actually be reinforcing the process of forest destruction. When protected areas have been established, for example, deforestation has often increased; and the introduction of state regulations on tree cutting have often led to forest areas being burned either because people have resented the fact that their livelihoods have been threatened or because they want to take advantage of loopholes which stipulate that trees that are burnt can be cut.

As we saw earlier in the discussion of communal forest protection in Guatemala, the introduction of state regulations on natural resource use can lead to a sort of institutional vacuum where a traditional system of communal or private regulation is displaced by another that remains ineffective. When the Mayan Biosphere Reserve was created in the Petén in 1990, the system of logging concessions was temporarily suspended. Deforestation, however, actually increased as logging companies began to purchase wood directly from peasants while some companies continued to cut wood illegally (Valenzuela, 1995). The same occurred in north-eastern Costa Rica when there was a rush to cut and sell timber before protected area status was effectively implemented in the Tortuguero region (Brüggemann, 1992).

The development of 'eco-capitalism' has also been subject to numerous constraints and contradictions. In most countries it lacks substance: many large companies actively publicize their new-found commitment to the environment but do little to change their practices on the ground. And what is often held up as sound corporate environmental policy or programme may actually have negative environmental and social impacts at the local level when, for example:

- primary or secondary forest is cleared to make way for monoculture tree plantations, or when forest plantations restrict people's access to forest products;
- the construction of infrastructure for eco-tourism is carried out in an unplanned way;
- 'eco-labelling' schemes that provide wood certificates to indicate that timber comes from companies practising sustainable forest management are based on dubious criteria and weak monitoring practices;
- 'bio-prospecting' in tropical rainforests for genetic resources amounts to 'bio-piracy', or another facet of a long historical process by which the North has appropriated resources from the South without due recompense, in this case both to the host country's government and to local populations that provide essential 'indigenous' knowledge.

What accounts for these constraints and contradictions? The simple answer is that environmental protection agencies often lack the resources necessary to implement programmes and enforce regulations. But there is also a more fundamental problem of approach. Until fairly recently, conservation in the region has been characterized by a fairly top-down technocratic approach to resource management.

This approach has been characterized by what can be called problems of 'macro- and microcoherency' (Utting, 1994). The former refers to the fact that governments are promoting environmental protection while at the same time implementing policies and programmes that encourage economic activities, or contribute to other pressures, that result in forest destruction. They may also serve to weaken institutions responsible for protection and rehabilitation of the environment. The latter refers to the fact that the technocratic top-down approach that characterises many environmental protection schemes often ignores the rights, needs, cultural practices and livelihood systems of local resource users. When this clash of interests occurs the local population is likely to respond in ways – involving apathy, noncooperation, conflict, sabotage and clandestine activities – that undermine efforts to arrest deforestation, let alone rehabilitate forest areas.

There are definite signs, however, that environmental or development agencies and practitioners throughout the region are engaged in an active learning process, identifying elements of success and failure, and experimenting with alternative approaches (Barzetti and Rovinski, 1992; Pasos *et al.*, 1994; Utting, 1993). This learning process is concerned primarily with overcoming the problem of 'micro-coherency' referred to earlier, *i.e.* promoting initiatives that respond positively to the livelihood concerns of local people. It is also about responding more to the ideas and demands of local people who often have their own plans and priorities for protecting the environment.

While this approach is clearly gaining in popularity and scope, it is still too early to tell whether it will effectively alter the dynamics of deforestation. Many of the innovations, methods and practices encouraged by this approach are applied in a piecemeal way and often assume the character of technical 'fixes' that do little to significantly alter the livelihood situation local resource users or to empower them. The mass of the indigenous population in Central America, and much of the peasant population in general, has experienced a history of repression, exploitation and marginalization that has created a culture of mistrust that affects relations with the institutions involved in forest protection and tree planting.

But perhaps the major constraint that limits the effectiveness of the participatory approach concerns the fact that it has made few inroads into the problem of macro-coherency mentioned above. The capacity of the participatory approach to promote sustainable development will always remain limited if policy makers continue to ignore the contradictory effects of government and donor agency policy. Several important structural inequities associated with the distribution of income and wealth are not being seriously addressed, yet it is these inequities that underpin social marginalization, disempowerment and many ongoing pressures on the natural resource base. Neo-liberal restructuring is a potent force in the region. Structural adjustment programmes are adding to pressures on the natural resource base, not least, as a result of increasing poverty. Economic stabilization programmes are imposing numerous constraints on environmental protection agencies. Moreover, there are signs that the neo-liberal trend towards the privatization of the state's functions may been extending from the economic to the environmental domain.[4] One feature of eco-capitalism has been the transfer of responsibility for environmental protection and regulation away from the public domain to the private sphere. The corporate sector is actively pushing for 'self-regulation' as opposed to external regulation by the state. But relying solely on the good faith of corporate management is not, perhaps, a very solid foundation for sustainable development.

Notes

1. This paper draws partly on information and analysis contained in Utting (1993) *Trees, People and Power: Social dimensions of deforestation and forest protection in Central America* (London: Earthscan).
2. The term *minifundismo* refers to rural settings where agricultural producers have access to extremely small plots of land, usually insufficient in size and quality to provide for the subsistence requirements of the family. *Minifundismo* is often associated with bimodal agrarian structures where small plots (*minifundia*) coexist locally or nationally with large estates (*latifundia*). In some areas of Central America, such as the western highlands of Guatemala, *minifundismo* has also been accentuated by intense population pressure and the sub-division of plots through inheritance.
3. This section draws on the results of an UNRISD field study, co-ordinated by Ileana Valenzuela, that formed part of a broader research programme on the social effects of deforestation in Central America that was co-ordinated by this author (see Valenzuela, 1991 and Utting, 1993).
4. This point has been made by Maria Antonieta Camacho in correspondence with this author.

Bibliography

Arias, O. and J. Nations (1992) 'A Call for Central American Peace Parks', in S. Annis (ed.), *Poverty, Natural Resources and Public Policy in Central America* (New Brunswick: Transaction Publishers).

Augelli, J. (1987) 'Costa Rica's Frontier Legacy', *The Geographical Review*, 77, 1–16.

Barzetti, V. and Y. Rovinski (eds) (1992) *Toward a Green Central America: Integrating Conservation and Development* (London: Kumarian Press).

Baumeister, E. (1991) *Elementos para Actualizar la Caracterization de la Agricultura Centroamericana*, Managua, mimeo.

Bonill Duran, A. (1988) *Crisis Ecologica en America Central* (San José: Ediciones Guayacan).

Bradley, T. *et al.* (1990a) *Guatemala Natural Resource Policy Inventory* (Vols I & II) USAID/ROCAP, Guatemala City.

Bradley, T. *et al.* (1990b) *Costa Rica Natural Resource Policy Inventory* (Vol. II) USAID/ROCAP, San José.

Brüggemann, J. and E. Salas Mandujano (1992) *Population Dynamics, Environmental Change and Development Processes in Costa Rica*, UNRISD, Geneva.

Cabrera Medaglia, J. (1995) 'Los contratos internacionales de uso de diversidad biologica', Humanismo y Medio Ambiente, special issue of *Praxis*, No. 49, 2/1995, pp. 223–9.

Carriere, J. (1991) 'The Crisis in Costa Rica: An ecological perspective' in D. Goodman and M. Redcliff (eds), *Environment and Development in Latin America: The politics of sustainability* (Manchester: Manchester University Press).

Centro de Derecho Ambiental y Recursos Naturales (CEDARENA) (1990) *Tortuguero Region Case Studies*, San José, mimeo.

Centro de Investigaciones y Estudios de la Reforma Agraria (CIERA) (1984) *Nicaragua: ... Y Por Eso Defendemos La Frontera*, CIERA-MIDINRA, Managua.

FAO (1995) *Production Yearbook* 1994, Vol. 48, FAO Statistics Series No. 125.

Heckadon Moreno, S. (1984) *Panama's Expanding Cattle Front: The Santeno Campesinos and the Colonization of the Forests*, University of Essex, Colchester.

Heckadon Moreno, S. and J. Espinosa (eds) (1985) *Agonia de la Naturaleza*, INDIAP/Smithonian Tropical Research Institute, Panama City.

Heckadon Moreno, S. and A. McKay (eds) (1982) *Colonizacion y Destruccion de Bosques en Panama* Asociacion Panamena de Antropologia, Panama City.

Houseal, B., C. MacFarland, G. Archibold and A. Chiari (1985) 'Indigenous Cultures and Protected Areas in Central America', *Cultural Survival Quarterly*, March, pp. 10–20.

Howard, P. (1987) *Frontier Expansion, Deforestation and Agrarian Social Change: The 'Cattleization' of the Nicaraguan Countryside, 1950–1971*, Universidad Nacional Autonoma de Honduras, Tegucigalpa, mimeo.

Instituto Nacional de Recursos Naturales Renovables (INRENARE) (1991) *Plan de Accion Forestal de Panama*, Documento Principal, Panama City.

Kaimowitz, D. (1995) *Livestock and Deforestation in Central America in the 1980s and 1990s: A Policy Perspective*, mimeo.

Kaimowitz, D. and O. Segura (1995) *The Political Dimension of Implementing Environmental Reform: Lessons from Costa Rica*, mimeo.

Leonard, H.J. (1987) *Natural Resources and Economic Development in Central America: A Regional Environmental Profile*, International Institute for Environment and Development (New Brunswick: Transaction Books).

Lutz, E., M. Vedova, H. Martinez, L. San Rarnon, R. Vasquez, A. Alvarado, L. Meiino, R. Colie, and J. Hnising (1993) *Interdisciplinary Fact-Finding on Current Deforestation in Costa Rica*, Environment Working Paper No. 61, Environment Department (Washington DC: World Bank).

Mansur, E. (1990) *El Salvador: Plan Nacional de Reforestacion de El Salvador*, FAO, San Salvador.

Ministerio de Agricultura y Ganadena (MAG) de Nicaragua (1990) *Incremento de la Productividad Agropecuaria y Conservacion de Recursos Suelo, Aguas y Bosques*, Managua: MAG.

Ministerio de Recursos Naturales, Energia y Minas de Costa Rica (MIRENEM) (1990) *Plan de Accion Forestal para Costa Rica: Documento Base*, San José: MIRENEM.

Nations, J. and D. Komer (1987) 'Rainforests and the Hamburger Society', *The Ecologist*, Vol. 17, No. 4/5.

Pasos *et al.* (1994) *El Ultimo Despale: La Frontera Agricola Centroamericana*, FUNDESCA, San José.

Plan de Accion Forestal de Guatemala (1990) *La Contribucion del Sector Forestal al Desarrollo*, Documento Tematico, Guatemalan Government, Guatemala City.

Reed, D. (ed.) (1992) *Structural Adjustment and the Environment* (London: Earthscan).

SECPLAN, DESFIL and USAID (1990) *Perfil Ambiental de Honduras 1989*, Tegucigalpa.

Silliman, J. (ed.) (1981) *Draft Environmental Profile of the Republic of Costa Rica*, Arid Lands Information Center, Tuscon.

Trejos, R. (1992) 'El comercio agropecuario extraregional', in C. Pomareda (ed.) *La agricultura en el desarrollo economico de Centroamerica en los 90*, San José, IICA.

Utting, P. (1993) *Trees, People and Power* (London: Earthscan).

Utting, P. (1994) 'Social and Political Dimensions of Environmental Protection in Central America', *Development and Change*, Vol. 25, No. 1, January.

Veblen, T. (1978) 'Forest Preservation in the Western Highlands of Guatemala', *The Geographical Review*, Vol. LXVIII, pp. 417–34.

Vivian, J. (1991) *Greening at the Grassroots: People's Participation in Sustainable Development*, UNRISD Discussion Paper No. 22, UNRISD, Geneva.

Weeks, J. (1985) *The Economies of Central America* (New York: Holmes & Meier).

Williams, R. (1986) *Export Agriculture and the Crisis in Central America* (Chapel Hill: University of North Carolina Press).

3 Macroeconomic Conditions for Sustainable Agriculture in Central America

Andy Thorpe[1]

1 INTRODUCTION

Agrarian policy discussions have traditionally centred on how policy can be formulated so as to galvanise agricultural production (the 'reversing of the terms of trade bias against agriculture' being one obvious facet) or to reduce rural inequalities (as epitomised by the various land redistribution laws passed in the sixties and seventies). In terms of agricultural policy objectives, sustainability of the resource base has generally trailed in last. This chapter redresses the balance, integrating the environmental issue in theoretic terms in Section 2, notwithstanding the difficulties of introducing further policy goals into an already complex policy framework, see Section 3. Section 4 then considers how the macro-economic environment may have facilitated environmental degradation through the extensification of agriculture and Section 5 through intensification and Section 6 discusses the externalities such farming developments have generated.

2 AGRICULTURAL PRODUCTION *AND* THE ENVIRONMENT OR *VERSUS* THE ENVIRONMENT?

'A central requirement for achieving better performance in the future as well as stabilising rural populations and reducing poverty, is that the large population of smallholders and landless labourers must have better access to the technology, inputs and services they need *to raise output and productivity'* (FAO, 1988, p. 16; the emphasis is ours). It is our contention that, in solely advocating the equity and efficiency aspects of agrarian policy, statements such as the FAO one above, carry an implicit presumption that unrestrained economic growth is compatible with longrun ecological sustainability. The reasoning, as Silva (1994, p. 698) notes, links poor economic performance to increasing poverty, which in turn fuels environmental degradation. Hence, the best way to reverse environmental degradation in the long-run is to promote growth, albeit at the expense of the environment, in the short-run. This can best be expressed diagrammatically (Figure 3.1). AE represents the transformation frontier between agricultural output and environmental services[2] for a given technology set (we assume a closed economy for simplicity). The elastic proportion of the curve ED reflects the fact that certain (low) levels of agricultural production are costless in environmental terms. Beyond point D however, increased agricultural output is only attainable through environmental sacrifice, hence the frontier is concave between AD. The precise nature of this trade-off will be dictated by a number of factors, these include (i) the structure of landholding, (ii) prevailing price vectors and (iii) producer behaviour.

The production equilibrium will depend on the country's social welfare function, SWF. These differ, as: 'In relatively high-income economies the income elasticity of demand for commodities and services related to sustenance is low and declines as incomes continue to rise, while the income elasticity of demand for more effective disposal of residuals and for environmental amenities is high and continues to rise' (Ruttan, 1971, pp. 70–8).

Hence the incorporation of environmental regulations into the SWF of high-income countries ensures that equilibrium outputs are biased towards D as opposed to A SWF[HY].[3] The reverse is true for low-income countries

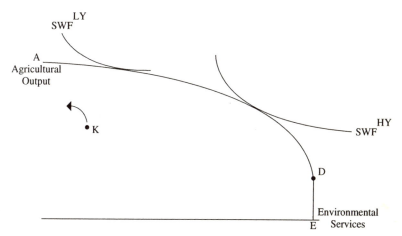

Figure 3.1   The trade-off between environmental services and agricultural output

SWF^{LY}. For a low-income country operating within AE (say at K), the appropriate expansion path will favour agricultural production at the expense of the environment. This in turn is perfectly compatible with political processes in which the government (i) refrains from implementing regulations which constrain income or employment-generating activities (Antle, 1993, p. 787) or (ii) has to be induced to adopt environmental initiatives through external aid conditionalities (Utting, 1994, p. 235).

Extending our model temporally offers the opportunity of technological change. In the long-term, the agricultural production/environmental services dichotomy is not a 'zero-sum' game and so new agricultural and conservation technologies can ensure either a restored or an improved production opportunity set becomes available (Lindarte and Benito, 1993, pp. 23–4). A restored production opportunity set sees the transformation frontier becoming AD^{1}E^1 in a two-period model (Figure 3.2). Now, for the former low-income economy, the 'dash for growth' in t = 0 has borne fruit and with the new prosperity has come a growing awareness of environmental issues (being reflected in a SWF that shifts in the direction of SWF^{HY}). Hence the production equilibrium in t = 1 will be closer to D.[1]

However what happens if environmental degradation is only partially reversible? In other words, there is an effective trade-off between production and the environment, a trade-off that means the environmental stock in t = 0 is greater than that bequeathed to society in t = 1. Agricultural output can still be maintained – but only by sacrificing further environmental capital (destruction of forest reserves, water and soil contamination). The

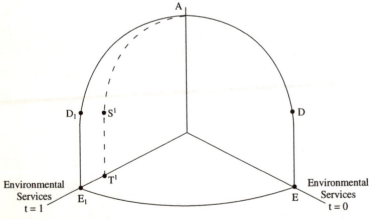

Figure 3.2   The agricultural output–environmental services trade-off over time

'degraded' transformation frontier AS^1T^1 is now inside AD^1E^1 and, as the temporal transformation surface $ADET^1S^1$ lies inside the sustainable $ADEE^1D^1$ then intertemporal welfare will almost certainly fall.

If a nation – or for that matter, an individual farmer – is cognizant[4] that environmental capital is being erased in an irreversible manner then the question must arise as to why it/she is prepared to condone such depletion? Trigo's response (1991, p. 19) is that these are rational outcomes given the prevailing development model. From a *macro-perspective*, a politician's time horizon doesn't generally extend further than the proximate elections. An optimal vote-maximisation on strategy turns a blind eye to environmentally harmful activities with long-term implications (such as deforestation, illegal colonisation and the use of prohibited pesticides) as the alternatives (redistribution of land, strict enforcement of existing – or the introduction of new – legislation) are costly when evaluated in electoral terms.

At a *micro-level*, the day to day struggle for survival requires the poor to intensify production. Soil over-exploitation is fine if the soil is able to recuperate its productive qualities. If not, then a vicious circle emerges as long-term costs are exchanged for short-term benefits, with the poor becoming poorer with the passing of the seasons. It is not just the poor who mortgage the land's future. Intertemporal wealth-maximisation by those who possess multiple income sources can have the same result.[5]

If outcomes are rational, yet trade-off between agricultural output growth and the environment makes the model inherently unstable, then we are faced with an acute dilemma: the maintenance of environmental capital in the long-run may be incompatible with current farming practices.

3 APPROPRIATE POLICY LEVELS – THE RELATIONSHIP BETWEEN THE MICRO AND MACRO

If current farming practices are indeed responsible for environmental degradation (an issue we turn our attention to in subsequent sections) then agrarian policy needs to be re-orientated, building in environmental safeguards. The issue is a complicated one however as the agrarian sphere is but one sub-set of the national economic environment, albeit an important one given the structure of the Central American economies. In recognising this, it follows that there are three different decision-making planes by which agrarian development is influenced: (i) the national, (ii) the sectoral and (iii) the local.

Trade liberalisation strategies are illustrative of the *national* dimension of policy vis-a-vis the agrarian sector. Liberalisation is expected to increase economic activity, with the main beneficiaries being regions such as Central America where (i) regulatory standards are more relaxed and (ii) production costs are low. This is as true for agriculture as it is for industry; Abler and Pick (1993) estimate that Mexican horticultural regions such as Sinaloa are likely to reap greater harvests, albeit at greater environmental cost, as a direct consequence of the NAFTA deal.

At a *sectoral* level, the nature of agrarian development can be influenced by governmental output (in particular, guaranteed producer prices and the establishment of procurement agencies), input (credit, feed, seed, fertiliser, and so on) and land (titling, redistribution, tenancy reform schemes, colonization programmes) policies. Historically, agrarian policy has had two distinct impacts upon the Central American environment. First, it has promoted the extension of the cultivated area, most obviously through deforestation, pushing forward the agricultural frontier and leading to the incorporation of new soils which may not be best suited to agricultural exploitation (see Section 4). Second, it has promoted the intensification of farming (Section 5). Whilst there is scope for the instigation of environmentally-sustainable agriculture projects and policies at the *local* level, this level is also the receptor of national and sectoral policy – in the sense that it is here that the reactions to, and repercussions of, macro-policy are felt.[6]

This is critical, for it would be an error to ignore the fact that optimal environmental strategies, while economically sound, may be politically unfeasible. Policy-makers are not immune from the pressures of powerful social groupings, interested transnationals or even local *caciques*; with the consequence that the path from policy prescription to policy adoption offers a host of opportunities for intended legislation to be subverted (Fandell, 1994; Silva, 1994, pp. 702ff; Valeriano, 1987). Hence, unless environmental protection policies are; (i) located within a coherent, comprehensive development policy framework, (ii) reflect prevailing political and institutional constraints and (iii) most critically, command local community support, the legislation has a limited probability of success. With these caveats in mind, we now turn to examine how macro-economic policy has contributed to environmental degradation.

4   THE EXTENSIFICATION OF FARMING – THE AGRICULTURE-FORESTRY DILEMMA

Deforestation is not a new phenomenon in Central America, however as land has become progressively more scarce, concern has been expressed over the rate at which forest resources are being felled by crop producers, livestock raisers and timber extractors (Kaimovitz, 1996; Utting, 1996, in this text). Of the 96.3 million hectares of hillside land – the most environmentally sensitive land-type – found in Central America, Mexico and the Tropical Andes, presently just 3.9 million hectares remains covered in natural vegetation. Deforestation of this terrain has resulted in 11.6 million hectares being seriously denuded of top-soil through water erosion, with a further 14.2 million hectares suffering moderate losses. The major soil losses have been in recent years (Jones, 1993). This degradation, which pushes society onto the inferior transformation surface ADEPT^{sQ1} (Figure 3.2), is conditional upon three factors:

Legal causes – the land tenure problem

In historic terms, as forested areas were an open access resource, then the revenue-maximising strategy was to exploit cleared land until the net returns generated were surpassed by the benefits of shifting production into a new stretch of virgin forest. This is all very well while the forest can absorb these expansionist pressures. The problem emerges when the forest becomes scarce; production is extended onto marginal soils, and tenure uncertainty prevents the establishment of sustainable-forestry and agricultural practices.

Cárcomo, Alwang and Norton (1994, p. 258) and Lindarte and Benito (1993, p. 76) stress that secure land titling is essential if peasants are to be encouraged to invest time and money in soil conservation technologies. Similarly Utting (1994, p. 243) points out that there is little incentive for an occupant without title to invest in a reafforestation exercise where the expected benefits mature in the long-term, and yet there is a strong likelihood of eviction in the short-term.[7] Tenure security is also perceived to offer the deed-holder improved access to formal credit channels, with production efficiency improving as a formal land market emerges (Stanfield, 1990, p. 164ff). In environmental terms, the removal of these credit and land market imperfections should lower peasant discount rates, making the adoption of sustainable farming practices more likely. In the case of forested areas two issues are of particular importance; (i) what is the most appropriate form of tenure? and (ii) how are historic (unregistered) rights to the resource to be incorporated within the tenure system proposed.

Tenure and a robust legal framework are essential prerequisites if sustainable forestry and agricultural practices are to be designed and successfully implemented. However as Stanley (1991, p. 763) recognises: 'Property regimes alone do not determine environmental outcomes.' Title by itself then isn't enough – the promotion of sustainable forestry and farming practices needs something extra.

Economic causes – the (dis)incentive problem

The buoyancy of international commodity prices in the post-war period, allied to reduced production costs, courtesy of the incentives (subsidised credit, fertilisers, etc.) on offer, made it economically profitable[8] to clear forested areas. The promotion of colonization projects on the agricultural frontier have contributed to regional deforestation (Enríquez, 1991; Edelman, 1993; Thorpe, 1993) while, at times, forestry legislation has also been its own worst enemy (Binswanger, 1991, p. 823).

Despite this accelerating deforestation, the maintenance of an incentive framework favouring agricultural production over forestry was justifiable on two grounds, one ecological and the other economic. The notion that the agricultural production/environmental services dichotomy was not a 'zero-sum' game and technological developments permitted the recovery of environmental capital, peripheralised the ecological debate. Economically, incentives had been justified in the context of the ISI development model being pursued in Latin America at the time. The industrial bias envisaged in the ISI programmes was to be ameliorated by compensatory agricultural policies – hence the use of subsidies and credits – the exact destination of these incentives being dictated by governmental priorities at the time.[9]

More recent recognition of the production/environment trade-off has promoted a re-evaluation of the existing incentive structures, and in particular the need to prevent the deforestation of marginal agricultural soils. An appropriate incentive structure that redresses the balance and favours the maintenance of forested areas must have as minimum elements: (i) provisions to ensure that the steward of the forest receives a higher proportion of the value-added in the sector (ii) the instigation of penalties and fines that deter infractions, thereby internalising possible ecological damage and (iii) reafforestation plans that are both easily verifiable and supervisable.

Institutional causes – the capacity problem

Even the best devised land tenure systems and planned incentive schemes ultimately rely upon institutions to police them. If institutions are unable (or unwilling) to carry out these duties, the prospects of introducing an effective environmental protection strategy are slim. We can identify three reasons for institutional failure: (i) poorly defined regulations which leave the institution with an unclear mandate as to how it is expected to act, (ii) poorly defined roles whereby responsibilities are shared between two or more institutions and (iii) institutional inadequacy. This institutional failure has led to alternative management and policy transmission mechanisms being considered. One avenue which is attracting great interest (see Bebbington *et al.*, 1993) is the development of links or collaborative programmes with *Non-Governmental Organisations* (NGOs). NGOs, with their greater expertise at the local level, are seen as providing a more cost-effective method of attaining government social and environmental goals. Four cautionary points need to be raised: (i) little consideration has been given to date as to what the appropriate long-term state-NGO relationship should be (Kaimovitz, 1993, p. 193), (ii) if NGOs are 'crowded-in', becoming substitutes for, as opposed to complements of, the state at the local level, their contribution to the conservation of environmental capital will be limited, (iii) there are wide differences in capabilities within the NGO-sector itself, and (iv) NGOs, due to the nature of their funding sources, plan in a manner (one year, five years, etc.) which doesn't always co-incide with optimal environmental protection strategies.

5   THE INTENSIFICATION OF FARMING – PRODUCING WHAT? AND HOW?

If we accept that there are limits to the expansion of cultivated area, then the increases in agricultural output that are needed to feed a growing world population must come through the intensification of farming. This intensification[10] can take two forms: (i) a switch in production systems to a more commercial (remunerative) crop or (ii) a switch in farming techniques intended to raise the yield of a designated crop or crops – both of which will affect the farmers environmental capital. An increase in the intensification of farming may also require a concomitant de increase in pesticide and herbicide dosages. Such effects are only partially internalised (viz: positive impacts on yield) and the question of externalities (viz: pollution of the water-table) is dealt with in Section 6.

Production systems: commercial versus subsistence crops

It is unrealistic to expect the choice of production system to remain static over time. Exogenous changes such as (i) the development of new markets and the collapse of old, (ii) the introduction of new land tenure regimes and (iii) macro-economic changes in relative prices (to name but three) all coalesce to determine which crop is planted. Bunch and López (1994, p. 12) for example note how wheat growing was aborted in San Martín, Guatemala as peasants could achieve better returns by growing cauliflowers and broccoli for the market. Perhaps one of the most sophisticated production systems developed in peasant communities however is that of the Almolongans in Guatemala: 'To avoid rain-damage to the crop, we sow onions in october, harvesting them in January. Potatoes are then planted and, shortly before they're harvested in April or May, we interplant cabbage or cauliflower. Having harvested the potatoes the rain makes the soil ready for carrots, beetroot, radish, lettuce, coriander.' (AVANSCO, 1994, pp. 12–13).

While such a production system may maximise producer revenues it is not entirely costless, as is the case with a number of the new Non-Traditional Agricultural Export Crop (NTAX)[11] production systems found in the region. Rosset (1993, p. 37) suggests the downside to adopting these systems include: greater risk, higher cost production, intensive agrochemical use and a greater susceptibility to pests. Despite this downside, the rapid growth in the introduction of these crop types has been spurred by a number of conducive macroeconomic developments.

First, the Caribbean Basin Initiative launched by the Reagan administration in the early 1980s established research entities, producer organisations and export promotion agencies in the countries of the region. With the institutional base in place, the second step saw the introduction of national legislation designed to increase NTX (including NTAX) production. Costa Rica was first off the mark when the 1984 *Ley para el Equilibrio Financiero del Sector Público* created export contracts (more popularly known as CATs) and abolished export taxes on NTX. In El Salvador, Congress passed the *Ley de Fomento de Exportaciones* in March 1986, Honduras followed with an identically named Law in 1987, while in Guatemala the *Ley de Fomento y Desarrollo de la Actividad Exportadora y de Maquila* arrived on the statute books in June 1989. Nicaragua's export promotion laws arrived in the 1990s (De la Ossa and Alonso, 1990). Finally, the dismemberment of the last vestiges of the old model, most notably the elimination of overvalued exchange-rates, gave an additional fillip to exporters. In strict economic terms, our view is that the scales have tilted too far. Neo-liberal orthodoxy may reign and *laissez-faire* may rule – except as far as exporters are concerned.

There may be an additional environmental reason for removing these export preferences. If there is a difference in the degree of soil erosion or degradation associated with certain crops, then these preferences may either encourage or discourage soil erosion (Cárcomo, Alwang and

Norton, 1994, p. 257). The central question then is: Which types of crop least facilitate soil erosion or degradation and thereby reduce the production-environment trade-off?

De Groot (1994, p. 3) quotes a WRI study on Costa Rican soil loss over the period 1970–89, deducing that per hectare soil loss under annual crops is eight times higher than that encountered either on land under permanent crops or pasture. AVANSCO (1994c, p. 38ff) discovered erosion levels on a plot sown with a NTAX crop, broccoli, in San José Pinula, Guatemala, to be almost twice those measured on a maize plot during the same 15-day time period. Cárcomo, Alwang and Norton (1994, p. 268) corroborate AVANSCO's conclusions, stating how soil conservation is enhanced as vegetable prices fall, vegetables causing greater erosion than other crops. Jansen (1995, p. 7) is a little more cautious, having discovered that basic grain cultivation is actually more environmentally damaging than export crops (principally coffee) on hillsides around El Nispero, Honduras. He concludes: 'there exists no simple relation between the production of export crops and environmental degradation.'

Macro-economic policy can be used in an effective environmental manner when specific crop production systems are unambiguously Pareto-inferior (one suspects cotton would presently score highly on this barometer) or Pareto-superior from a national perspective. Discriminatory credits, inputs[12] and guaranteed prices can all be bestowed on producers in an effort to get them switching into/out of these identified production systems. The problem arises when a clear Pareto choice is indeterminate. In these cases, resolution is only likely to be effective at a regional or local level, taking account of location-specific constraints and alternatives.

Farming techniques – low or high external inputs?

Land use patterns, even for the same agricultural commodity, are in a state of continual flux. Johannessen (1963) traces how the removal of Crown regulations in 1800, which had forced cattle-ranchers to sell their stock at annual fairs, stimulated a cattle boom. As herds expanded and savannas became overcrowded, ranchers responded by fencing the commons. More intensive animal husbandry technologies saw man-made pastures being planted from about 1910 onwards with the high-yielding Jaragua grass, imported from Brazil by the Honduran Ministry of Agriculture in 1918, distributed freely to cattle-ranchers. Not all enterprises had either the land or the capital to survive as the market economy developed, and the same author recounts how the destruction of the grassland resource was paralleled by the collapse of the *hacienda* system in a number of regions.

De Groot (1994, pp. 4ff) acknowledges that peasants too are not immune from the pressures of the market economy. The growth of commercial agriculture not only introduced new crop production systems with consequent environmental implications, but altered the production techniques employed within peasant agriculture. Forced to abandon shifting agriculture in favour of more intensive systems of land use,[13] this shift was not without cost. Browning (1971, p. 234) quotes a 1954 Salvadorean soil study that recommended taking more than 50 per cent of peasant subsistence lands out of production if soil degradation was to be avoided.

Land scarcity generally leads to a shortening of the fallow period (Buckles and Arteaga, 1993, p. 52); to high peasant discount rates, biasing welfare functions towards the present and increasing the risk of the peasant located on the degraded transformation frontier ADET[1]S[1] (Figure 3.2).

Faced with a degraded soil resource, and the impossibility of rotation farming, an alternative appears to be the technification of production. Bebbington (1993, p. 64) notes how the poor have adopted agro-chemicals as a means of ameliorating the effects of commercialisation and land subdivision, and so it does not seem implausible to assume there could be a technical solution to the problem of land degradation. To assess this we need to resolve two questions. First, are high input techniques capable of restoring or improving the production opportunity set (the ecological question)? Second, if so, at what cost (the economic question)?

First, while the 'rational' use of inputs can restore or improve yields, excessive use of pesticides, herbicides or fungicides[14] can alter the acidity of the soil/change its composition in a manner which adversely affects soil productivity over the long-run. Pesticide use can also detrimentally affect harvests while leaving soil productivity unchanged in instances where insect immunity increases. Faber (1993, pp. 93ff) notes how the 1971 cohort of Nicaraguan cotton boll-weevils were found to be 45 times more resistant to methyl parathion than any previous generation.

It was not just cotton that had a pest problem. AVANSCO (1994a, pp. 15ff) note that while the tomato was the second most important crop cultivated in Eastern Guatemala in the sixties and-seventies, the growing immunity of the white-fly to the chemicals applied, had made production unprofitable in the region by the mid-eighties. Pesticide usage can also result in unacceptably high toxic residues in plants, residuals which result in lower commodity prices. AVANSCO (1994c, p. 1) note that in the first six months of 1992, the US rejected 296 shipments of Guatemalan agricultural commodities worth an estimated $9.6 million for this very reason.

Second, in answering the economic question, two qualifications must be made; (i) technified production is much more sensitive to prevailing factor and commodity prices. Hence while the preferred solution to boll-weevil infestation was 'more of the same', sprayings increasing from 5–10 (1950s) to 25–28 (1960s) to over 50 times (1970s and 1980s); this was partly prompted by the cheapness of the pesticides, (ii) technified production has higher 'up-front' production costs. As regional capital markets are imperfect (Noé Pino, Thorpe and Restrepo, 1995), access to agro-chemicals is not scale-neutral but is accordingly concentrated in the large farm sector.[15]

In the new macro-economic environment: 'Adjustment programmes have led to the removal of some of the price distorting mechanisms that had kept nominal prices for agrochemicals and imported machinery below real prices and that at the same time had subsidised the credit used' (Bebbington, 1993, p. 68). In a sense, if we accept that there are strong ecological reasons for rationalising agro-chemical use, the adjustment package, by changing input prices so as to reflect true social costs, is the appropriate policy action as regards the medium and large-farm sector.[16] The problem is rather more complex in the small farm sector. Here degradation occurs as the peasant over-exploits his plot. Now, by removing or reducing input subsidies, we eliminate a potential escape avenue – the question is (i) what alternative, ecologically favourable farming techniques exist, (ii) are these potentially adoptable by small peasants and (iii) how do we wean such peasants into introducing them?

While agronomists have identified a number of ecologically favourable soil conservation techniques (López-Pereira *et al.*, 1994; Ellis-Jones and Sims, 1995) and entomologists have delineated similarly natural or biological forms of pest control (Bentley, 1989) there has only been a limited diffusion of such methods within Central America.[17] The reasons for low take-up, according to farmers interviewed by AVANSCO (1994c, p. 42) in the Guatemalan Altiplano, include: (i) an ignorance of such techniques, (ii) the techniques available were either too labour or capital-intensive and (iii) as tenant-farmers it was irrational to undertake such strategies, for the proprietor refused to recognise the *mejoras* when the tenancy agreement ceased. The third of these grievances can only be resolved by tenure reform, while the second can be eliminated through the manipulation of the incentive structure, altering relative factor prices in a manner conducive to promoting conservation techniques. The need to wean peasants into introducing new techniques is part and parcel of the same need to increase peasant awareness of alternative techniques.

The dispersion of knowledge has a macro-level dimension in the sense that policies can facilitate experimentation and diffusion. It can also help identify key areas where soil degradation or pesticide overuse is acute, allowing resources to be targeted at these areas. However the 'agents of change' charged with raising producer awareness of alternative production techniques need to work at a more local or community level. Here there is more space for state-NGO collaboration, though the ultimate aim should be the creation of paratechnics in the local community itself. For this reason, the most effective soil conservation and pesticide policies in the long-run are likely to be micro-based, and (i) be easy to demonstrate (allowing farmers to train farmers), (ii) develop or improve upon existing practices and (iii) be socially and culturally acceptable.[18]

6   EXTERNALITIES, ECONOMICS AND THE ENVIRONMENT

Externalities arise where the production or consumption of a commodity either imposes a cost on third parties for which they remain un-compensated, or offers benefits to another grouping without the pro-ducer/consumer receiving recompense. In the context of this chapter externalities are generated by both the extensification and intensification of agriculture.

The most commonly cited relate to high external input agriculture and refer to the loss of human or animal life through agro-chemical misuse. Murray (1991, p. 22) cites the case of the poisoning of melon workers in Choluteca, Honduras who, after spreading pesticide by hand, promptly ate from the selfsame (unwashed) hands. Rosset (1993, p. 39) suggests that this is not unexceptional in the melon-fields. An AVANSCO (1994a, p. 51) investigation in Guatemala's Motagua valley found 39 per cent of agrochemical applications were undertaken without any form of protective clothing, 18 per cent of the sample admitting to not washing after completing the applications. In Guatemala 88 cases of pesticide poisoning are reported each month, in Costa Rica the figure is over 540 a month, while mothers in the Nicaraguan cotton regions have been found to have DDT levels in breast milk 42 to 185 times the recommended safe level (Faber, 1993, pp. 100ff). The same author also blames agrochemicals for the region's malaria epidemic.[19] Another study, this time by Aguilar (cited in AHE-FOPRIDEH, 1989, p. 22), attributes 24.4 per cent of the unexpected cattle deaths in Honduras over the period 1983–8 to agrochemical poisoning.

They are not the only externalities however. While the consequences of soil erosion are largely internalised via reduced output, the eroded soil can cause sedimentation build-up in rivers and so both reduce navigability and increase the risk of flooding. On a somewhat different tack, AVANSCO (1994c, p.

34) discovered that the incidence of insect pests in maize fields is greater, the closer the field is to agro-chemical intensive horticultural production – a finding they attribute to the destructive nature of pesticides on the local predator community.[20] In Guatemala too, a forest-protection scheme which encouraged the local community to thin forest areas generated an unintended externality as outsiders entered the thinned forest areas to illegally strip bark (Utting, 1994, p. 241). Not all externalities are negative; the extensification of agriculture generates positive externalities as previously isolated peasant groupings now benefit from improved market access due to the establishment of new road networks.

While the generator of externalities remains unaffected, society is pushed onto the inferior transformation frontier $ADET^1S^1$. To prevent this economists recommend that externalities arising from market failure, as in the instances cited above, should be resolved through the assignment of property rights. These property rights may; (i) outlaw certain activities completely (such as the proscription of DDT as an agricultural input in a number of the countries), (ii) permit particular activities, subject to a number of controls and regulations (the best example being the insistence on licensing forestry management in the private sector), (iii) tax (or subsidise) specific activities. The object of this intervention is to equate the marginal social benefits with the costs, thereby achieving a socially efficient allocation. The key policy question is: 'whether public interests warrant [such] an intervention [as]... intervention seems more justifiable if the social cost of soil conservation is high and private cost low' (De Groot, 1994, p. 6). The same argument applies not just to soil conservation, but can be extended to all externalities generated by the extensification and/or intensification of Central American agriculture.

7 CONCLUDING COMMENTS

The growth of extensive cultivation methods in Central America can be historically linked to inadequately developed legal frameworks, economic incentives, institutional failure and cultural attitudes. But, as the land frontier is pared away, intensive farming practices have grown in importance. Production as a consequence is now much more sensitive to movements in factor and commodity prices, irrespective of whether it is techniques or production systems that have changed. This does not mean individual farmers are as flexible in their production decision however. The individual production decision remains constrained by the land endowment, back-up capital and labour availability matrix of each cultivator. These in turn need to be mapped onto the payback, risk, income-generating characteristics, input requirements and forgone opportunity costs matrices associated with different crops. The result will be a wide array of potential production decisions from any given change in factor or commodity prices; wider still if the matrices are extended to incorporate differences in (i) long and short-run costs due to growing immunization of pests and (ii) returns due to soil exhaustion, as this article has shown.

It is for this reason that we concur with a number of authors (Utting, 1994, p. 245; Persson and Munasinghe, 1995, p. 259) who point out the difficulty of generalising over the impact macro-economic reforms are having on the environment. This notwithstanding, it is essential we analyze the trade-offs between the different policy objectives, searching for policies that minimise these trade-offs. In this sense we do not advocate reversing macro-economic policies designed to fulfil more pressing national objectives (unless the ecological impact of such policies at a national level are considered to be too adverse). Instead the goal should be to supplement them with complementary local or regional initiatives (as discussed in some of the following chapters) which cancel out the potentially ecologically harmful micro-responses.[21]

Notes

1. Thanks are offered to (i) members of the workshop on 'Sustainable Agriculture in Central America' that took place at the ASERCCA Conference in Paris, 13–14 October 1995 and (ii) Chris Reid and Terri Sarch of the University of Portsmouth for comments on earlier drafts of this paper.
2. Environmental services is an epithet for constant soil productivity, water and air quality, availability of and access to national parks and reserves.
3. As the agricultural sector in high-income countries is relatively small, the negative impact of these new regulations can be more easily offset by compensation schemes (Trigo, Kaimovitz and Flores, 1991, p. 16). The same option does not exist in low income countries where there is a dominant agricultural sector (Harold and Runge, 1993, p. 790).
4. Cognizance cannot be assumed. Ayres (1993, pp. 2ff) suggests that there is still controversy over whether any environmental capital sources have suffered irreparable damage. Furthermore, at the micro-level, Preston (1995, p. 3) notes that few farmers responding to a questionnaire in Tarija, Bolivia mentioned environmental degradation as a problem – with the exception of those who had assisted at either NGO or government courses on environmental problems.
5. The most obvious case is that of the fruit multinationals who can switch production to other areas (or countries) if land productivity falls.
6. Consideration of micro-level policies are beyond the remit of this particular chapter, being dealt with in some of the accompanying chapters. It is our

belief however that macro-policies alone cannot safeguard the environment but, given the heterogeneity of the countryside, must be complemented by micro and meso measures.

7. De Groot (personal communique) also points out that while titling may improve resource steward-ship in the long-term, it can increase deforestation rates in the short-term as people seek to establish de facto ownership rights (and hence eligibility to formal title) through forest clearance.

8. Those such as Ruhl (1984, pp. 39ff) and Williams (1986, p. 124) who support the thesis that the cattle boom squeezed peasants off fertile lands in the valleys, forcing them to become hillside farmers, would argue that the clearance of such land was more of an economic necessity.

9. For example, Abler and Pick (1993, p. 795) calculate that effective Mexican subsidies were around 40 per cent for fertilizer, 50 per cent for pesticides and 80 per cent for irrigation, with government agricultural credits on offer at negative real interest rates. AVANSCO (1993, p. 13) detail how 80 per cent of agrarian credit dispensed in Guatemala during the eighties was lent for crop production. A further 17 per cent went to cattle ranchers, while just 3 per cent ended up in the forestry, hunting and fishing sub-sector.

10. In the context of this paper intensification refers to the desire to increase revenues and/or yields from a given land area. It will not be used to describe increases in labour or capital productivities, although there is likely to be a positive correlation between the variables.

11. We define NTAX crops as those that do not appear in the country's export statistics in a regular or significant manner prior to the 1980s and 1990s Structural Adjustment programmes.

12. This is presuming the substitutability of both inputs and credits between different crops is low.

13. It was not just land that was harvested more intensively. AHE-FOPRIDEH (1989, p. 22) note that water resources became more intensively mined as dynamite replaced more conventional fishing gear.

14. Most emphasis is focused on pesticide overuse and so we shall concentrate on this aspect unless otherwise stated.

15. The 1993 Honduran Agricultural Census (Volume VI, p. 19) corroborates this. 49.9 per cent of farms under 5 hectares use agro-chemicals, compared to 60.3 per cent of those with 5–10 hectares and 66.5 per cent of those with 10 hectares plus.

16. The actual impact of this policy on input purchases is unknown, but will be crucially dependent upon demand elasticities for the respective agro-chemical. No surveys to date have attempted to estimate this, though anecdotal evidence suggests input demand may be highly inelastic.

17. The recent Honduran Agricultural Census (1993, p. 24) for example discovered that just 4 per cent of smallholders had introduced techniques to conserve soils (the figures being 7.1 per cent and 10.1 per cent for farmers with 5–10 and over 10 hectares respectively.

18. We stress these three aspects as numerous studies have shown that projects which introduce new, complex techniques have a life-span that runs little further than the end of project. Although projects reflecting these criteria are not guaranteed success, the participative element fostered by such projects gives them a greater chance of taking root (Bunch and López, 1994).

19. A recent Internet item from InterPress (26/8/1995) commented on the number of hemorraghic dengue cases in the region. With the government in El Salvador intent on combatting the epidemic through the fumigation of the mosquitoes habitat, a leading environmentalist in the country commented: 'the remedy could be worse than the cause.'
20. This externality is internalised if the neighbouring field is owned by the horiculturalist. If not, it remains an externality.
21. For example, devaluation is introduced to encourage exports and discourage imports, thereby rectifying trade imbalances. If agro-export production now increases and these production systems have negative environmental implications (as with the greater erosion caused by broccoli crops in Guatemala) the appropriate policy response is not to reverse policy and revalue the currency. Rather, a regional initiative to promote soil conservation techniques through barriers/walls etc. would correct the problem while allowing the devaluation to cure the macro-economic ill.

Bibliography

Abler, D.G., Pick, D. (1993) 'NAFTA, Agriculture, and the Environment in Mexico', *American Journal of Agricultural Economics*, Vol. 75, pp. 794–8.

AHE-COHDEFOR (1994) *El Sistema Nacional de Areas Protegidas de Honduras*, Tegucigalpa.

AVANSCO (1994) *Agricultura Intensiva y Cambios en la Comunidad de Almolonga, Quetzaltenango*, Texto para Debate No. 2., Guatemala.

AVANSCO (1994a) *Apostando al Futuro con los Cultivos No-Tradicionales de Exportación* I, Texto para Debate No. 3., Guatemala.

AVANSCO (1994c) *Impacto Ecológico de los Cultivos Hortícolas No-Tradicionales en el Altiplano de Guatemala*, Texto para Debate No. 5., Guatemala.

AVANSCO (1993) *Agricultura y Campesinado en Guatemala*, Texto para Debate No. 1, Guatemala.

Ayres, R.U. (1993) *Eco-Restructuring: The Transition to an Ecologically Sustainable Economy*, INSEAD Centre for the Management of Environmental Resources WP 93/95/EPS.

Bebbington, A., G. Thiele, P. Davis, M. Prager, H. Riveros (1993) *Non-Governmental Organizations and the State in Latin America* (London: Routledge).

Bentley, J.W. (1989) 'What Farmers don't Know can't Help Them: The Strengths and Weaknesses of Indigenous Technical Knowledge in Honduras', *Agriculture and Human Values*, Vol. 6(3), pp. 25–31.

Binswanger, H.P. (1991) 'Brazilian Policies that Encourage Deforestation in the Amazon', *World Development*, Vol. 19(7), pp. 821–9.

Browning, D. (1971) *El Salvador: Landscape and Society* (Oxford: Clarendon Press).

Buckles, D., L. Arteaga (1993) 'Extensión Campesino a Campesino de los Abonos Verdes en la Sierra de Santa Marta, Veracruz, México', in D. Buckles (ed.) *Gorras y Sombreros: Caminos hacia la Colaboración entre Técnicos y Campesinos*, Memoria de Taller, Veracruz.

Bunch, R., V.G. López (1994) *Soil Recuperation in Central America: Measuring the Impact Three to Forty Years after Intervention*, Paper presented at the International Institute for Environment and Development's International Policy Workship, Bangalore, India, 28 November–2 December.

Cárcomo, J.A., J. Alwang, G.W. Norton (1994) 'On-site Economic Evaluation of Soil Conservation Practices in Honduras', *Agricultural Economics*, Vol. 11, pp. 257–69.

Edelman, M. (1993) 'Illegal Renting of Agrarian Reform Lands: A Costa Rican Case Study' in Glade, W., C.A. Reilly, *Inquiry at the Grassroots*, Inter-American Foundation, Arlington, Virginia.

Ellis-Jones, J., B. Sims (1995) 'An Appraisal of Soil Conservation Technologies on Hillside Farms in Honduras, Mexico and Nicaragua', *Project Appraisal*, Vol. 10(2), pp. 125–34.

Enríquez, L. (1991) *Harvesting Change: Labour and Agrarian Reform in Nicaragua, 1979–90* (Chapel Hill: University of North Carolina Press).

Faber, D. (1993) *Environment under Fire: Imperialism and Ecological Crisis in Central America* (New Y: Monthly Review Press).

Fandell, S.E. (1994) 'Foreign Investment, Logging, and Environmentalism in Developing Countries: Implications of Stone Container Corporation's Experience in Honduras', *Harvard International Law Review*, Vol. 35(2), pp. 499–533.

FAO (1988) *Potential for Agricultural and Rural Development in Latin America and the Caribbean*, Annex V: Crops, Livestock, Fisheries and Forestry, Rome.

Groot, J.P. de (1994) *Policies for Soil Conservation*, CDR Documento #94001, San José.

Harold, C., C.F. Runge (1993) 'GATT and The Environment: Policy Research Needs', *American Journal of Agricultural Economics*, Vol. 75, pp. 789–93.

Honduras, *Agricultural Census* (1993, 1994), Various Volumes.

Interpress (1995) *El Salvador: Experts balme Dengue Epidemic on Environmental Imbalance*, 26/8/1995.

Jansen, K. (1995) 'Sistema de Maíz y Café y el Cambio Ecológico en las Montanas de Santa Bárbara, Honduras', *Cambio Ecologico en Honduras: Sus Vinculaciones con la Producción, la Comercialización, el Estado y las Organizaciones de Desarrollo*, POSCAE Documento de Trabajo No. 9, Tegucigalpa.

Johannessen, C.L. (1963) *Savannas of Interior Honduras* (Los Angeles: University of California Press).

Jones (1993) *Hillsides Definition and Classification*, mimeo.

Kaimovitz, D. (1996) *Livestock and Deforestation in Central America in the 1980s and 1990s: A Policy Perspective*, mimeo.

Kaimovitz, D. (1993) 'NGO's, the State and Agriculture in Central America' in A. Bebbington, G. Thiele, P. Davis, M. Prager, H. Riveros, *Non-Governmental Organizations and the State in Latin America* (London: Routledge).

López-Pereira, M.A., J.H. Sanders, T.G. Baker, P.V. Preckel (1994) 'Economics of Erosion-Control and Seed Fertilizer Technologies for Hillside Farming in Honduras', *Agricultural Economics*, Vol. 11, pp. 271–88.

Lindarte, E., C. Benito (1993) *Sostenibilidad y Agricultura de Laderas en América Central: Cambio Tecnológico y Cambio Institucional*, IICA Serie Documentos de Programas No. 33, San José.

Murray, D.L. (1991) 'Export Agriculture, Ecological Disruption, and Social Inequity: Some Effect of Pesticides in Southern Honduras', *Agriculture and Human Values*, Fall, pp. 19–29.

Noé Pino, H., A. Thorpe, A.L. Restrepo (1995) *El Crédito Agrícola en el Ambiente Centroamericano* (POSCAE-UNAH, Tegucigalpa).

Ossa, A. de la, E. Alonso (1990) *Exportaciones no Tradicionales en Centroamérica*, Cuardernos de Ciencias Sociales No. 31. (FLACSO, San José).

Persson, A., M. Munasinghe (1995) 'Natural Resource Management and Economywide Policies in Costa Rica: A Computable General Equilibrium (CGE) Modeling Approach', *World Bank Economic Review*, Vol. 9(2), pp. 259–85.

Preston, D. (1995) 'Crisis Ecológica. Para Quien?', *Cambio Ecologico en Honduras: Sus Vinculaciones con la Producción, la Comercialización, el Estado y las Organizaciones de Desarrollo*, POSCAE Documento de Trabajo No. 9, Tegucigalpa.

Rosset, P. (1993) 'El Manejo Integrado de Plagas (MIP) y la Producción Campesina de Cultivos No-Tradicionales', *Revista de CLADES*, Vol. 5/6, pp. 36–41.

Ruhl, J.M. (1984) 'Agrarian Structure and Political Stability in Honduras', *Journal of Interamerican Studies and World Affairs*, Vol. 26(1), pp. 33–68.

Ruttan, V.W. (1971) 'Technology and the Environment', *American Journal of Agricultural Economics*, Vol. 53, pp. 707–17.

Silva, E. (1994) 'Thinking Politically about Sustainable Development in the Tropical Forests of Latin America', *Development and Change*, Vol. 25, pp. 697–721.

Stanfield, D. (1990) *Titulación de la Tierra, Seguridad de Tenencia y Desarrollo Rural en Honduras*, Tegucigalpa, mimeo.

Stanley, D.L. (1991) 'Communal Forest Management: The Honduras Resin Tappers,' *Development and Change*, Vol. 22, pp. 757–79.

Thorpe, A. (1993) *Land Markets in Honduras*, Tegucigalpa, mimeo.

Trigo, E.J. (1991) *Hacia una Estrategia para un Desarrollo Agropecuario Sostenible* (IICA, San José).

Trigo, E., D. Kaimovitz, R. Flores (1991) *Toward a Working Agenda for Sustainable Development*, IICA Program Papers Series No. 25, San José.

Utting, P. (1996) *Deforestation in Central America: Historical and Contemporary Dynamics*, mimeo.

Utting, P. (1994) 'Social and Political Dimensions of Environmental Protection in Central America', *Development and Change*, Vol. 25, pp. 231–59.

Valeriano, E.F. (1987) *La Explotación Bananera en Honduras*, UNAH Colección Realidad Nacional No. 17, Tegucigalpa.

Williams, R.G. (1986) *Export Agriculture and the Crisis in Central America* (Chapel Hill: University of North Carolina Press).

Part II

Sustainable Production Systems

4 Policies Affecting Deforestation for Cattle in Central America

David Kaimowitz[1]

1 INTRODUCTION

The most important change in land use in Central America in the last 40 years has been the widespread conversion of forest to pasture. Between 1950 and 1990, the area in forest in the region fell from 29 million hectares to 17 million, and the majority of cleared lands became pastures (Utting, 1992).

Some land clearing was justified. But in many cases, the costs of deforestation outweighed the benefits (Ledec, 1992). Large amounts of wood and non-timber forest products were wasted. Soil degradation and siltation increased. Genetic resources were lost. Carbon dioxide released by burning forests contributed to global warming. Moreover, many new pastures can only sustain their nutritional value for cattle for a few years under current practices.

The benefits of future deforestation will probably be even lower and the costs higher. The remaining forests tend to be on marginal lands with excessive rains and poor soils or steep slopes. Cattle raising or crop production in these areas generates low levels of income per hectare, but the forests generally have high levels of biodiversity and are important for watershed protection.

The literature provides seven explanations for why pasture has expanded at the expense of forest:

- Favorable markets for livestock products (Myers, 1981).
- Government subsidies for livestock credit and road construction (Binswanger, 1991).
- Land tenure policies that promote deforestation to establish property rights (Jones, 1990).

51

- Slow technological change in livestock that favors extensive production systems (Serrao and Toledo, 1993).
- Policies which depress timber values and make forest management unprofitable (Kishor and Constantino, 1993; Stewart and Gibson, 1994).
- Reduced violence, which has lowered the risk of ranching in isolated areas (Maldidier, 1993).
- Characteristics of cattle such as their low labor and supervision requirements, transportability, limited risk, prestige value, limited use of purchased inputs, and biological and economic flexibility (Hecht, 1992).

Depending on which factors ones believes to be more important, the prognosis and policy recommendations emerging from the analysis are different. This study uses the Central American experience during the last fifteen years to put forth some hypotheses about how the seven factors listed above have influenced the conversion of forest to pasture and about how effective policies designed to address these issues have been or are likely to be.

The study covers all of Central America, except E1 Salvador, which has little remaining natural forest and where changes in forest cover have been only marginally related to trends in the livestock sector in recent years.

The study concludes that falling beef and dairy prices only moderately reduce the extent of forest clearing for pasture and affect cattle population and pasture area in traditional cattle grazing regions more than on the agricultural frontier. Technological change in cattle raising and elimination of policies which lower timber prices are also unlikely to reduce forest clearing, since their supposed effect on deforestation comes through similar relative price shifts as those associated with changes in beef and dairy prices. Besides, technological changes in cattle raising in Central America probably cannot influence beef prices, which are largely determined on the international market, and higher timber prices can increase pressure for timber removal from unmanaged forests. On the other hand, changes in road construction, land tenure, and land use policies could greatly reduce forest clearing for pasture, although they are unlikely to eliminate it entirely. Livestock credit is not currently a major cause of forest clearing, but should be restricted in agricultural frontier regions with high rainfall.

Deforestation rates in Central America as a whole in the late 1980s were lower than ten years earlier, but continued to be high and in some places are again rising, due to the end of military conflicts, public support for road construction in forest regions, and the increasing political power of cattle ranchers.

The study has ten sections. The second looks at what has happened with land use and cattle in Central America over the last fifteen years. The third describes the major types of livestock ranches in Central America and their relative importance. Then come six sections which analyze how each of the factors listed above has affected the observed tendencies and the final conclusions.[2]

2 FORESTS, CATTLE, AND PASTURES

Recent estimates show a total deforestation rate for Central America of between 324 000 and 431 000 hectares per year (Table 4.1) The majority of deforested land has been transformed into pastures (Ledec, 1992). Nevertheless, several of the deforestation estimates either overestimate deforestation during the period covered or are no longer applicable, since deforestation has declined since the original studies were made. After reviewing the available evidence, this author estimates that total deforestation in Central America diminished from around 400 000 hectares per year in the late 1970s to about 300 000 hectares in 1990. Deforestation declined in Costa Rica, Nicaragua (during the 1980s), and Panama. On the other hand, it increased in Petén, Guatemala and Nicaragua (since 1990).

Table 4.1   Recent estimates of annual deforestation (thousands of hectares) in Central America (excluding El Salvador)[*]

	Grainger	*Nations &* *Komer*	*WRI*	*FAO*	*Merlet*	*Utting*
	(76–80)	*(82)*	*(81–85)*	*(81–90)*	*(91)*	*(90)*
Costa Rica	60	60	50	50	40	50
Guatemala	na	60	90	81	90	90
Honduras	53	70	90	112	108	80
Nicaragua	97	100	121	124	125	70
Panama	31	50	36	64	41	34
Total	na	340	402	431	394	324

[*] The years in parenthesis are the years for which the figures supposedly apply. However, all the figures are based on studies carried out in the mid-1980s or earlier.
Sources: FAO, 1993; Grainger, 1993; Merlet *et al.*, 1993; Nations and Komer, 1983; Utting, 1993, WRI, 1992.

Table 4.2   Cattle Population in Central America in 1950, 1970, 1978 and 1992
(excluding El Salvador) (million head)

	1950	*1970*	*1978*	*1992*
Costa Rica	0.6	1.5	2.0	1.7
Guatemala	1.0	1.5	2.1	2.2
Honduras	0.9	1.2	1.8	2.1
Nicaragua	1.1	2.2	2.5	2.2
Panama	1.1	1.2	1.4	1.4
Total	0.6	7.6	10.1	9.6

Sources: For 1950 all countries: Leonard (1987). For other years: Costa Rica:
1970 and 1978: FAO (1980), 1993: Consejo Nacional de Producion, unpublished
data; Guatemla: 1970 and 1978: Banco de Guatemala (1981), 1992: Banco de
Guatemala, unpublished data; Honduras: 1970 and 1978: based on extrapolations
from the 1965 and 1974 censuses, 1992: SECPLAN (1994); Nicaragua: Holmann
(1993); Panama: Dirección de Estadística y Censo (1992).

At the same, time between 1950 and 1978, the region's cattle herd more
than doubled (Table 4.2). After that it stagnated, and in 1992 the region
had fewer cattle than fourteen years earlier. For the most part, changes in
national pasture areas followed closely the changes in cattle population.

The national statistics hide major differences among regions within
each country. Cattle and pastures have tended to expand in the humid
Atlantic plains, but in the traditional cattle producing areas of the Pacific
and interior they have tended to decline (Merlet, 1992).

The decline in cattle in traditional cattle grazing regions has led to a
major increase in abandoned lands, which have become brush and even
secondary forest. This process in strongest in Nicaragua and Costa Rica,
but is also occurring in other countries.

3   THE DIFFERENT 'LOGICS' OF LIVESTOCK PRODUCTION

The Central American livestock sector has distinct types of producers,
who respond differently to changes in policy, markets, and technology.
For purposes of this study, four major types of livestock producers have
been identified: (i) 'traditional' medium and large ranchers, (ii) 'investment' ranchers, (iii) medium and large ranchers on the agricultural frontier, and (iv) small ranchers.

Traditional medium and large ranchers come from families that have been in the cattle business since before 1950. Much of their land was inherited or obtained at minimal cost by making claims on public lands. These lands have opportunity costs, but do not involve any cash outlays, and many of the ranchers do not include them in their calculations when assessing the profitability of their operations.

Investment ranchers, in comparison, are capitalist entrepreneurs with little experience with cattle raising, who have purchased land and cattle in recent decades because it appeared to be a profitable investment. They tend to base their decisions more on short term profit margins (including the opportunity costs of land and cattle), have low 'barriers to exit' from cattle production, and sell their land or reduce their cattle stock when business is poor.

The medium and large agricultural frontier ranchers can be distinguished from the first two groups by their physical residence in agricultural frontier areas. For ranchers on the agricultural frontier, cattle offer the critical advantage of being easy to transport. In some cases, they even walk to market themselves. Moreover, once annual crop yields begin to fall due to declining fertility and weed infestation, conversion to pasture is often the only economically viable use for frontier land (Hecht, 1992).

There are also many small ranchers, since the first thing that most small farmers in Central America do if they accumulate a little land or money is to purchase cattle. The limited availability of family labor constrains the expansion of crop production, and farmers prefer to avoid the cash outlays and supervision time required to hire large amounts of outside labor. Cattle also have the additional advantages for small farmers of being a convenient form of low-risk and easily convertible savings, providing regular income from the sale of dairy products, and making use of marginal or degraded lands which can no longer sustain crops (Hecht, 1992).

Policy instruments which operate through livestock and forest product prices are more likely to influence the land use patterns of investment ranchers and of ranchers in traditional livestock grazing areas than of ranchers who live on the agricultural frontier. The latter groups tends to have low supply elasticities for their livestock production, because they have few profitable alternatives to invest their savings in.

4 THE ROLE OF MARKET FORCES

In 1981, Myers coined the term 'the hamburger connection' to describe how the expanding US market for Central American beef generated a cattle boom, which in turn led to widespread deforestation. During the beef export boom associated with this phenomena, Central American beef exports rose from 9 million dollars in 1961 to 290 million dollars in 1979 (Williams, 1986).

Since the mid-1970s, however, the outlook for Central American beef exports has worsened. International beef prices have been low, the US has imposed protectionist measures which limit Central America's access to the US market, and various government policies have tended to depress real beef prices (Cajina, 1986; Edelman, 1985; Howard, 1987). As a result, between 1978 and 1985, Central American beef exports declined from 120 million metric tons to only 49 million tons (Torres-Rivas, 1989).

Due to the declining profitability of beef exports fewer new investors went into cattle after 1979, except for specific situations where rising land prices and available subsidized credit made ranching profitable for other reasons. When the price declines began, many traditional medium and large ranchers retained their cattle to wait for higher prices. But when prices did not improve, they were often forced to sell their cattle at even lower prices to pay their debts (Howard, 1987). Many heavily indebted ranchers and smaller ranchers with minimal liquidity were forced out of business entirely (Maldidier, 1993; Van der Weide, 1986). On the other hand, the influence of price changes on many smaller and more isolated ranchers was limited, since these ranchers had few viable livelihood alternatives and medium and large farmers continued to purchase new farms on the agricultural frontier to take advantage of the comparatively low land prices there (Hijfte, 1989).

5   SUBSIDIZED CREDIT AND PUBLIC ROAD CONSTRUCTION

During the beef export boom in the 1960s and early 1970s, the amount of livestock credit in real terms and the percentage of agricultural credit allocated to livestock grew rapidly in all the countries. Livestock credit during this period was heavily subsidized through both below market interest rates and leniency with respect to loan recuperation and was allocated to a relatively small group of ranchers.

Subsidized credit for cattle promoted deforestation in several ways (Ledec, 1992b). Credit helped ranchers to overcome capital constraints, which would have otherwise limited pasture expansion. Large ranchers used a significant amount of livestock credit directly to purchase lands which they might otherwise have been unable to afford. Credit subsidies for livestock made cattle a more attractive investment prospect compared to other alternatives.

Nevertheless, the role of subsidized public credit in the conversion of forest to pasture should not be exaggerated. Ledec (1992b) shows that only about seven to ten percent of deforestation in Panama could be attributed to public livestock credit and notes that in general banks prefer to lend to large established ranchers in traditional cattle raising areas, rather than to the generally poorer ranchers along the agricultural frontier. These conclusions are also supported by the fact that both the Peten Guatemala and Eastern Honduras have experienced major pasture expansion even though relatively little public livestock credit has been available in those areas.

In the last fifteen years, public livestock credit has become less available and less subsidized. Real livestock lending in Guatemala peaked in 1973 and then declined through 1989 (Vargas *et al.*, 1991). In Costa Rica, it began to decline in 1981, and by 1989 had fallen to the same level as in 1970 (Holmann *et al.*, 1992). Lending in Honduras, Nicaragua, and Panama rose until the second half of the 1980s but then fell abruptly (Maldidier, 1993; Sarmiento, 1992; Ventura, 1992).

Even so, the only country where reduced access to subsidized credit has had a major impact on cattle production is Nicaragua, where most ranchers have major liquidity problems after years of political turmoil and economic crisis (Maldidier, 1993). The decline in credit availability has only had a moderate effect in the other countries, which has probably been greater in traditional cattle grazing regions than along the agricultural frontier.

On the other hand, if there is one single government policy which has had a major and indisputable impact on promoting conversion of forest to pasture it is road construction in forested regions (AHT-APESA, 1992; Jones, 1990; Ledec, 1992). This is reflected in a 1987 statistical analysis of deforestation in northern Honduras, which found that 'those areas nearest to roads are most susceptible to deforestation. The further an area is from a road, the smaller the percentage of (area) affected by deforestation. Beyond five kilometers there was a rapid drop in the percentage cleared' (Ludeke, 1987, p. 76). In Panama, 'colonization and eventual deforestation are likely to occur within 2–10 kilometers of either side of an all-weather rural road which has penetrated a frontier area. This implies a deforestation area of influence of 400 to 2000 hectares for each new kilometer of road built in forested zones' (Ledec, 1992, p. 199).

By providing access to new areas road construction makes it easier to enter and deforest and cheaper to transport cattle and dairy products from the area. Roads also stimulate land speculation. The magnitude of forest destruction that road building has caused can easily be understood if one considers that between 1953 and 1978 the length of all-weather roads in Central America rose from 8 350 kilometers to 26 700 kilometers (Williams, 1986).

Unlike favorable market conditions and credit subsidies which tended to disappear in the 1980s, public road construction in forested areas continued largely unabated. Annual growth in roads in Costa Rica increased from 6.5 per cent between 1974 and 1980 to 10.4 per cent between 1981 and 1990 (Holmann *et al.*, 1992). Recent road construction in Panama, Guatemala, and Nicaragua also brought many colonists to previously forested areas.

6 LAND TENURE POLICIES AND LAND MARKETS

Purchasing or claiming lands in Central America has traditionally been quite profitable, and in most agricultural frontier areas real land prices tend to rise over time (Colchester and Lohmann; Edelman, 1985). One reason for this is that road construction has made many lands more accessible. Some authors also point to a purely speculative component to land prices which 'involves people attributing to deforested land an asset value that is well in excess of its actual production value' (Ledec, 1992, p. 96).

The rise in land prices has been so significant in agricultural frontier areas, that a large percentage of pasture expansion may have as much or more to do with land speculation as with cattle raising per se (Banco de Guatemala, 1981; Van der Weide, 1986). 'In some cases families just have cattle on land to show ownership while they speculate with land prices. Land ownership is a long-term investment, which is more important than the income from livestock as such' (Hijfte, 1989, p. 16).

Most land converted from forest to pasture over the last few decades originally belonged to the government. These lands could be legally claimed by farmers if they could show that they had been occupying them for more than a certain number of years. Often the laws required colonists to clear the forest in order to acquire possession rights, and in some cases permitted farmers to obtain title for larger amounts of land if it was for pasture than if it were for crops. And even when laws do not specifically require deforestation to demonstrate land possession, land clearing and the subsequent planting of pasture has still been one of the best ways to discourage squatters and avoid the threat of agrarian reform action designed to put 'idle lands' into use (Edelman, 1992; Place, 1981).

With the end of redistributive agrarian reform policies in Central America in the 1990s and recent modifications in the titling legislation designed to eliminate some of the incentives for deforestation, the incentive to convert forest to pasture to ensure tenure security may have weakened (Richards, 1994; Utting, 1992). Nevertheless, these changes in titling legislation often still have little effect on the behavior of government agencies at the local level and in many agricultural frontier areas land clearing remains the only effective mechanism for claiming possession (Hernández-Mora, 1994; Silviagro, 1992). Some governments also still actively promote land settlement through agricultural colonization programs.

On the other hand, not all government land policies have promoted forest clearing. Nicaragua's agrarian reform, for example, which began in 1979, was an important deterrent to private investment in cattle ranching and offered tens of thousands of families an alternative for gaining access to land different from moving to the agricultural frontier (Jarquin and Videa, 1990). Similarly, the creation of national parks and other protected areas in Costa Rica has also discouraged the conversion of forest to pasture. There has also been growing interest in providing tenure rights for indigenous people. It is argued that 'Indian populations tend to be less destructive of natural resources than others groups in the region' (Herlihy, 1992); and while this may not always be true, to date most indigenous groups in Central America have cleared little forest for pasture and are unlikely to do so in the near future.

7 TECHNOLOGICAL CHANGE AND ENVIRONMENTAL DEGRADATION

Pasture researchers in Latin America have long argued that technological improvements in livestock production systems can reduce pressure on marginal lands on the agricultural frontier by making it possible to produce the same amount of meat and milk on less land (Serrao and Toledo, 1993). This process is supposed to work through market mechanisms: As the efficiency of cattle production on existing pastures increases, the price of meat and milk will fall by more than the productivity gains and this will lower the incentive to put marginal lands into pasture.

There is now sufficient technology available to double or perhaps even triple average stocking densities in Central America. Even if this were to occur, however, it is unlikely that it would lead to a decline in deforestation. The experience of the last twenty years shows that pasture expansion in agricultural frontier areas can continue despite major declines in real beef and dairy prices. This is so because, as shown above, land speculation is often as important a factor in pasture expansion in these areas as the profitability of livestock itself, pastures are often the lowest cost land use which provides tenure security, and farmers in frontier areas typically have few viable production alternatives other than cattle. Moreover, in the current context of increasing trade liberalization it is far from clear that changes in the efficiency of regional livestock production will affect prices, which are now determined mostly in the world market. A plausible argument can even be made that improved livestock technology applicable to areas with poor soils in the humid tropics is likely to increase deforestation, as it would make cattle raising in these areas more profitable.

8   FORESTRY POLICY

Another view of why farmers have cleared forest land in Central America is that government policies have lowered the value of forest land and forest products, and hence the potential profitability of maintaining the land in forest (Kishor and Constantino, 1993; Stewart and Gibson, 1994). These authors argue that the use of log export bans, low public expenditure on forestry; restrictions on cutting timber, and cumbersome requirements for forestry management plans have discriminated against the forestry sector and made forestry less profitable compared to cattle raising and crop production.

Stewart and Gibson estimate that if there were no policy distortions both management of native forests and forest plantations would currently yield higher returns per hectare of land than cattle in Costa Rica; and they imply that one simply has to remove these distortions and ranchers will stop deforesting and start planting trees (1994). Kishor and Constantino, on the other hand, affirm that even without trade distortions cattle raising would still provide higher incomes than continuous management of natural forests, although they agree with Stewart and Gibson that forestry plantations would be more profitable than ranching (1993).

The basic argument that policies which lower the value of timber discourage reforestation and secondary forest regeneration is undoubtedly valid. Nevertheless, in addition to comparing the net present value of cattle versus forestry, policy analysts must also consider the other reasons why landowners have preferred cattle over other investment options such as limited labor requirements, low supervision costs, ease of sale on short notice, and the advantages of cattle as a way to demonstrate land possession. Different types of forestry management have some of these attributes, but not necessarily all of them. Probably the only group of ranchers for which net present value per hectare is the overriding factor in defining land use are the investment ranchers. This implies that improvements in timber prices alone are unlikely to eliminate all conversion of forest to pasture, although they may reduce it.

There has also been a confusion in much of the literature referring to natural forest management between incentives which will promote sustainable management of natural forests, secondary forest regrowth, and new forest plantations and those that may encourage rapid extraction of valuable timber, thus leaving a forest of greatly diminished commercial valuable which would be subject to clearing. Policies which favor reforestation and secondary forest regeneration by increasing timber prices and giving individual clear property rights over timber may at the same time encourage land clearing of primary forests.

9 POLITICAL INSTABILITY AND VIOLENCE

The Nicaraguan civil war during the late 1970s reduced the national cattle herd by between 25 per cent and 35 per cent (Cajina, 1986). This was followed by further anti-government violence between 1983 and 1989, which again discouraged cattle production and brought the expansion of the agricultural frontier to a virtual standstill (Maldidier, 1993).

The formal end of the military conflict in 1990 brought with it the return of thousands of displaced families to the agricultural frontier, but has not meant a complete end to rural violence. Land markets in the interior became more active in 1990 and 1991 after the war ended, but then became virtually paralyzed again, due to continued problems of insecurity. Frequent assaults and kidnappings and ranchers' inability to take effective possession of lands they purchase have continued to make it unappealing to buy land or locate cattle in inaccessible areas (Matus et al., 1993).

Violence in Guatemala also had a major impact on the cattle sector, particularly in the Peten and Alta Verapaz. After 1980, growing violence in the Peten associated with Guatemala's internal military conflict substantially reduced the number of ranchers interested in purchasing land from the government and led many ranchers to abandon their lands or to sell them for low prices. This situation continued until 1987 or 1988, after which the level of violence declined.

Had it not been for the military conflict in the 1980s deforestation in Guatemala and Nicaragua would have probably been much higher than it was. Since the levels of violence have subsided in these countries, investments in cattle and land prices have increased rapidly.

10 CONCLUSIONS

While deforestation is still a major problem in Central America, it declined significantly in the late 1980s. Unfavorable market conditions for meat and dairy products, reduced access to credit for livestock, higher interest rates, expansion of protected areas, and military conflicts all contributed to this decline. And while there is little hard evidence on the subject, the changes in market conditions and credit probably had a larger impact on the behavior of investment ranchers and other large ranchers outside agricultural frontier areas than it did on pasture expansion in agricultural frontier areas.

Forest clearing for pastures over the last fifteen years can largely be attributed to attempts to claim public lands and improve land tenure security, public road construction, government colonization programs, and the specific characteristics of cattle raising which make it quite attractive to farmers with land but little capital, labor, management skills, and access to markets.

These conclusions are also consistent with recent experience in the Brazilian Amazon (Hecht, 1992; Moran, 1993). There also, deforestation rates came down in the late 1980s and early 1990s as a result of the termination of credit and fiscal subsidies and economic recession. But after initially falling, deforestation in Brazil seems to have reached a plateau, with most of the land clearing being carried out by small and medium sized farmers who are less affected by changes in government subsidies or the price of livestock products.

To bring deforestation rates down farther, whether it be in Central America or other areas of tropical Latin America, will require going beyond the elimination of subsidized public credit and fiscal incentives for cattle ranching. And given what has been shown about the limited responsive of forest clearing to changes in livestock product prices, trying to artificially lower the prices of livestock products through consumer boycotts or other measures is also unlikely to do the job. Stimulating improved livestock technology and eliminating policies which depress timber prices have a number of virtues in their own right, but will probably not significantly reduce the clearing of natural forest, and could even increase it.

To further reduce deforestation in Central America below its current level will require directly addressing the issues of road construction, land tenure, land use regulation. Road construction and improvement in most forest areas should be discouraged. Governments must decide which public lands they do not want to pass into private hands and strictly enforce those decisions, and the incentives must be eliminated for clearing forests to claim land and improve tenure security. These governments cannot be realistically expected to maintain control over all current public lands, but they should attempt to keep control of prioritized areas. Protected areas must really be protected, while at the same time establishing the best possible relations with neighboring communities. Indigenous land rights should be expanded and cattle ranching should be restricted in buffer zones around protected areas.

Notes

1. The author is grateful for comments by Gerardo Budowski, Neil Byron, Ronnie de Camino, Jan de Groot, Marc Edelman, Sam Fujisaka, Alicia Grimes, Peter Hazell, Federico Holmann, Elizabeth Katz, Manuel Paveri Anziani, Sara Scherr, Denise Stanley, William Sunderlin, Lori Ann Thrupp, and Steven Vosti. Support for the research was provided by Deutsche Gesellschaft für Technische Zusammenarbeit (GTZ), Inter-American Institute for Cooperation in Agriculture (IICA), International Food Policy Research Institute (IFPRI), United States Agency for International Development (USAID), and the World Bank. The author, however, is solely responsible for the material presented.
2. The discussion of the specific characteristics of livestock has been incorporated into the section on the types of livestock ranches.

Bibliography

Agrar-Und Hydrotechnick – Asesoría y Promoción Económica S.A. (AHT-APESA), (1992) *Plan de Desarrollo Integrado del Petén*, Convenio Gobiernos Alemania / Guatemala, Vol. I, Diagnóstico General del Petén.

Banco de Guatemala (1981) 'Establecimiento de una empresa ganadera en el Petén', *Informe Económico*, 28, April–June, 1981, pp. 21–70.

Binswanger, H.P. (1991) 'Brazilian Policies that Encourage Deforestation in the Amazon', *World Development*, Vol. 19, No. 7, pp. 821–9.

Cajina, A. (1986) *Ganadería bovina en Nicaragua, Recuento crítico y retos del presente*, Cuaderno de Investigación no 4, Managua: INIES.

Colchester, M. and L. Lohmann, *The Struggle for Land and the Fate of the Forests* (London: Zed Books).

Dirección de Estadística y Censo (1992) *Estadística panameña, Situación económica, Producción pecuaria, año* 1990, Panama.

Edelman, M. (1992) *The Logic of the Latifundio, The Large Estates of Northwestern Costa Rica Since the Late Nineteenth Century* (Stanford: Stanford University Press).

Edelman, M. (1985) 'Extensive Land Use and the Logic of the Latifundio: A Case Study in Guanacaste Province, Costa Rica', *Human Ecology*, Vol. 13, No. 2, pp. 153–85.

Food and Agriculture Organization of the United Nations (FAO) (1993) *Forest Resources Assessment 1990, Tropical Countries*, FAO Forestry Paper 112, Rome: FAO.

Food and Agriculture Organization of the United Nations (1966, 1976, 1979, 1990, 1991) *Production Yearbooks*, Rome: FAO.

Grainger, A. (1993) 'Rates of Deforestation in the Humid Tropics: Estimates and Measurements', *The Geographical Journal*, Vol. 159, No. 1, March, pp. 33–44.

Hecht, S.B. (1992) 'Logics of Livestock and Deforestation: The Case of Amazonia', in Downing, T., S. Hecht, H. Pearson, C. García Downing (eds) *Development or Destruction, The Conversion of tropical Forest to Pasture in Latin America* (Boulder: Westview Press).

Herlihy, P.H. (1992) 'Wildlands Conservation in Central America During the 1980s: A Geographical Perspective', *Conference of Latin Americanist Geographers*, Vol. 17/18, pp. 31–43.

Hernández-Mora, N. (1994) *Effects of Policy Reform on Land Use Decisions and Community Forest Management in Honduras: Four Case Studies*, Masters Thesis Cornell University.

Hijfte, P.A van, (1989) *La ganadería de carne en el norte de la zona atlántica de Costa Rica*, CATIE /Wageningen/MAG, field report No. 31.

Holmann, F., R. Dario Estrada, F. Romero, L. Villegas (1992) 'Technology Adoption and Competetiveness in Small Milk Producing Farms in Costa Rica: A Case Study', presented at the Animal Production Systems Global Workshop, September 15–20, San Jose, Costa Rica, 1992.

Holmann, F. (1993) *Costos de producción de leche y carne, inversión de capital y competitividad en fincas de doble propósito en cinco regiones de Nicaragua*, Managua: Comisión Nacional de Ganadería.

Howard, P. (1987) *From Banana Republic to Cattle Republic: Agrarian Roots of the Crisis in Honduras*, PhD dissertation, Madison: University of Wisconsin.

Huising, J. (1993) *Land Use Zones and Land Use Patterns in the Atlantic Zone* of Costa Rica, Masters Thesis, University of Wageningen.

Jarquin, J. and L.M. Videa (1990) *Los sistemas de producción ganaderos en la V región y el impacto de las políticas económicas hacia el sector*, Departamento de Economía Agrícola, UNAN, Managua.

Jones, J.R. (1990) *Colonization and Environment, Land Settlement Projects in Central America* (Tokyo: United Nations University Press).

Kishor, N.M. and L.F. Constantino (1993) *Forest Management and Competing Land Uses: An Economic Analysis for Costa Rica*, LATEN Dissemination Note #7, Washington DC: World Bank.

Ledec, G. (1992) 'New Directions for Livestock Policy: an Environmental Perspective', in Downing, T., S. Hecht, H. Pearson, C. García Downing (eds), *Development or Destruction, The Conversion of Tropical Forest to Pasture in Latin America* (Boulder: Westview Press).

Ledec, G. (1992b) *The Role of Bank Credit for Cattle Raising in Financing Tropical Deforestation: An Economic Case Study from Panama*, PhD dissertation, University of California, Berkeley.

Leonard, J. (1987) *Recursos naturales y desarrollo económico en América Central: un perfil ambiental*, Turrialba: CATIE.

Ludeke, A.K. (1987) *Natural and Cultural Physical Determinants of Anthropogenic Deforestation in the Cordillera Nombre de Dios, Honduras*, PhD dissertation, Texas A & M University.

Maldidier, C. (1993) *Tendencias actuales de la frontera agrícola en Nicaragua*, Nitlapan-UCA, Managua.

Matus, J., M. Padilla and F. Diaz (1993) *Diagnóstico de Campo*, Volume III, Nicaragua, Programa de Apoyo al Fortalecimiento de la Situación de Derecho y el Despegue Económico en el Campo, Informe de la Misión de Identificación, European Economic Community, Managua.

Merlet, M. (1992) *La region interiure et la frontiere agricole*, unpublished manuscript.

Merlet, M., G. Farrell, J. Laurent, and C. Borge (1992) *Identification de un programa regional de desarrollo sostenible en el tropico humedo, informe de consultoria*, Groupe de Recherche et d'Echanges Technologiques (GRET), Paris.

Moran, E. (1993) 'Deforestation and Land Use in the Brazilian Amazon', *Human Ecology*, 21:1, pp. 1–21.

Myers, N. (1981) 'The Hamburger Connection: How Central America's Forests Became North America's Hamburgers', *Ambio* 10:1, pp. 3–8.

Nations, J. and D. Komer (1983) 'Rainforests and the Hamburger Society', *Environment* 25:3, pp. 12–20.

Place, S. (1981) *Ecological and Social Consequences of Export Beef Production in Guanacaste Province, Costa Rica*, PhD dissertation, University of California, Los Angeles.

Richards, M. (1994) *Case Studies from Honduras, Central America*, draft.

Sarmiento, M. (1992) *Situación actual y perspectiva de la producción de la leche en Panamá*, Ministerio de Desarrollo Agropecuario / Instituto de Investigación Agropecuaria de Panama / Centro Internacio-nal de Investigación para el Desarrollo, Panama.

Secretaría de Planificación, Coordinación y Presupuesto (SECPLAN) / Secretaría de Recursos Naturales (SRN) (1994) *IV Censo Nacional Agropecuario*, Tegucigalpa.

Serrao, E.A. and J.M. Toledo (1993) 'The Search for Sustainability in Amazonian Pastures', in Anderson, A.B. (ed). *Alternatives to Deforestation: Steps Toward Sustainable Use of the Amazon Rain Forest* (New York: Columbia University Press).

Silviagro S. de R.L. (1994) *Analisis del sub-sector forestal de Honduras*, Tegucigalpa.

Stewart, R. and D. Gibson (1994) *Environmental and economic Development Consequences of Forest and Agricultural Sector Policies in Latin America (A Synthesis of Case Studies of Costa Rica, Ecuador, and Bolivia)*, draft.

Torres-Rivas, E. (1989) 'Perspectivas de la economía agroexportadora en Centroamerica', in Pelupessy, W., (ed.) *La economía agroexportadora en Centroamerica: crecimiento y adversidad*, San Facultad Latinoamericana de Ciencias Sociales, San José.

Utting, P. (1992) *Trees, People and Power, Social Dimensions of Deforestation and Forest Protection in Central America* United Nations Research Institute for Social Development, Geneva.

Van der Weide, P.A. (1986) *Exploratory Survey in the Atlantic Zone of Costa Rica, Animal Production*, MAG/Wageningen/CATIE, Field Reports No. 3.

Vargas, H., R. Dario Estrada, E. Fernando Navas, (1991) *Desarrollo de la ganadería bovina de doble propósito en parcelamientos de la Costa Sur de Guatemala*, Proyecto Mejoramiento de Sistemas de Producción Bovina de Doble Propósito en Guatemala, San José.

Ventura, F. (1992) *Diagnóstico Agroindustrial de la Carne de Ganado Bovino, Periódo 1980–1991*, INCAE, Tegucigalpa.

Williams, R. (1986) *Export Agriculture and the Crisis in Central America* (Chapel Hill: University of North Carolina Press).

World Resources Institute (1992) *World Resources 92–93, A Guide to the Global Environment*, WRI, New York.

5 Production Systems in the Humid Tropics of Nicaragua: a Comparison of Two Colonization Areas

Jan P. de Groot, Rosario Ambrogui and
Mario Lópes Jiménez[1]

1 INTRODUCTION

One important mechanism that could be exploited to conserve the tropical rain forest in Central America is to try to influence the colonization process, by reducing the flow of migrants to the agrarian frontier. Policies to achieve this involve giving peasants in the regions of outmigration access to land, technology, investment funds and markets. However, the progress made in these areas is limited. The alternative approach is to make colonist land use systems more productive and more ecologically sustainable. This approach can only succeed if the more sustainable practices are appropriate for the technical and socio-economic conditions of the colonists. More needs to be known about production systems in colonization areas in order to be able to design, adapt and transfer such practices. It is just as important to know the farmer's actual knowledge and perceptions relating to innovative practices.

This chapter based on a survey in two colonization areas in the humid tropics of Nicaragua in 1995 involving 105 farmers, aims at analysing the production systems of farmers in these areas and their knowledge and perceptions of alternative agricultural practices that are both more productive and more sustainable, in order to explore the possible role of such practices in farmers' development strategies.

Section 2 discusses general aspects of colonization in Central America: push and pull factors of colonization, limitations of farming systems in the humid tropics, and options to enhance sustainable production practices. In Sections 3 and 4 two settlements are compared, one in the older colonization area of Nicaragua, the other in the recent agrarian frontier. The comparison refers to stages of development, production systems, farm development strategies, and the conditions favouring the adoption of alternative production practices.

2   GENERAL ASPECTS OF COLONIZATION IN CENTRAL AMERICA

Push and pull factors of colonization

The push factors of colonization in Central America have been discussed extensively: the market incorporation of agriculture, the modernization of export production, cattle ranching, and the related changes in peasant subsistence agriculture (Williams, 1986; Utting, 1991; Howard-Borjas, 1992; and Kaimowitz, 1995). Peasants are increasingly drawn into the market economy and forced to rely on intensification of their land use systems, or to migrate to the agrarian frontier.

Cattle ranching which expanded in the 1960s, in response to the opening of the US market for imported beef, has been a main push factor. Whereas export crops displaced peasants only from the better land, ranching came to occupy land of every agricultural potential (Williams, 1986, p. 4; Howard-Borjas, 1992, p. 2). Moreover, the extremely low labour/land ratio in cattle ranching has exacerbated rural unemployment, provoking additional migration to the agrarian frontier.

The pull factors of colonization to the humid tropics are various (Bakker, 1993, p. 18). Colonization is an escape for unemployed, land poor and landless peasants. Extraction activities (hardwood, rubber, mining) attract labour, which, when these projects end, often turns to subsistence agriculture. Official colonization schemes have settled groups of rural and urban poor, but often have also attracted spontaneous colonists to project areas. Road building, credit programmes and other facilities have further induced migration (Kaimowitz, 1995, pp. 34–9).

Limitations of farming systems in the humid tropics

An increasing number of farmers in the humid tropics are faced with land degradation; as the vegetative cover decreases, the productive capacity of the land declines. Land degradation is a consequence of a reduction in the available biomass and of soil degradation; often the water regime is also affected (IFAD, 1992, p. 4). Soil degradation in the humid tropics is mainly caused by nutrient depletion, a decline in soil fertility due to agricultural practices. In humid and semi-humid climates where vegetation tends to be ample, the main storehouse of nutrients is not the soil but the standing vegetation. Certain soils cannot store nutrients and organic matter well. If the vegetation is eliminated, such soils will be exhausted after one or two crops (Ruthenberg, 1976, p. 44).

The traditional system of colonization of the rain forest is shifting cultivation. After cultivating one or two crops on a cleared plot, a forest fallow is observed that permits the regeneration of a secondary forest. At the end of the fallow period the plot is again cleared for agriculture. This system is usually ecologically stable and sustainable (Ruthenberg, 1976, p. 45). Colonists in Central America from the interior region of the isthmus or from the Pacific Coast are neither familiar with the ecosystem of the rain forest, nor do they operate – as in traditional rain forest farming systems – a closed economic system.

Socio-economic factors are therefore important in explaining the prevailing land use systems in colonization areas. The colonist is partly integrated into the market economy, but most markets are incomplete or imperfect. Arriving without any capital, the peasant initiates an extractive type of agriculture with relatively high yields, at low inputs. But this is only temporary. The rapid depletion of soil fertility forces him to cut the remaining forest, whose products contribute little to his income. As problems of low soil fertility and weeds are quick to arrive, and commercial crops face uncertain markets, there is little scope for capital accumulation. Pressing subsistence needs keep the farmer's rate of time preference high; as a consequence investment in restoring soil fertility or weed control are not perceived as beneficial.

One way to escape this low productivity trap is by repeated migration (Bakker, 1993, p. 23). As the value of land in the colonization area increases, the colonist can sell land which he has sown to pasture, and migrate to the next, the new, agrarian frontier. By selling land the colonist acquires working capital, which improves his starting position for capital accumulation in a new area. Ultimately, capital is not invested in land, but in cattle. Various authors have described this mechanism (for Central America see Maldidier *et al.*, 1993; Pasos, 1994; Kaimowitz, 1995).

Cattle ranching is a logical outcome of agricultural activities in Central and South American rain forest areas (Bakker, 1993, p. 19). After clearing the forest, and after two or three subsistence crops the soil is over-exploited and its fertility is depleted, and when left fallow it is invaded by natural pastures of low nutritional value, or the farmer sows pastures. Cattle ranching in colonization areas is also the consequence of imperfect or missing markets for a diversified crop production. Cattle can be transported relatively easily over large distances, even when infrastructure is poor. Consequently, cattle raising becomes the logical option for the colonist. The peasant either sells the land to a rancher wanting to extend his pastures, or he starts building up his own herd, emphasizing accumulation of cattle instead of land.

Cattle ranching has a low productivity, as factors of production are used extensively over a long period of time (Howard-Borjas, 1992, p. 3). Pasture degradation is the main problem facing cattle raising in the humid tropics. Pasture productivity declines because the low soil fertility is an important constraint to grass species; only less demanding species of low nutritional value and poor palatability will grow. Cattle ranchers also face problems of weed invasion.

If soil fertility, insect attack, diseases and weeds are not a problem for grass species and management is appropriate, it is possible to maintain pastures at a low level of productivity and at a sustainable equilibrium. In practice, however, grasslands commonly degrade rapidly, weed species invade and biomass continues to decline. Production per unit of land decreases; the average stocking rate of cattle can decline from one animal per ha. to one animal per five ha (Bakker, 1993, p. 20). If degradation proceeds this can also lead to complete abandonment of the land, which then starts to revert to a secondary forest type of vegetation.

Options to enhance sustainable practices in the humid tropics

The reduction of migration flows *to* the agrarian frontier is crucial in diminishing the pressure on the rain forest. At the same time, programmes *within* the colonization areas to enhance productivity and sustainability of the farming systems of colonists in their present settlements are required. Such programmes aim at turning around the downward spiral of the continuing extraction economy, including the re-migration to the new agrarian frontier. Such programmes can only succeed if they adapt to the existing farming systems of the colonists. In the older colonization areas the social differentiation has advanced more than in the more recent agrarian frontier. Different categories of farmers can make use of different innovations.

Examples of more sustainable practices include: cover crops, home gardens, alley cropping, pasture reclamation and silvopastoral systems. Most of the more sustainable practices are tree-based: tree-pasture or tree-crop combinations. In general these systems require initial investments and are relatively complex to manage. Their adoption has consequences for the entire farming system: for tillage practices, soil fertility management, weed and disease control, labour allocation. They have to be fine-tuned at the farm level and will only work if farmers learn to manage, to adapt and to transfer them.

3 COMPARING THE OLDER AND THE RECENT AGRARIAN FRONTIER

On its eastern coast Nicaragua possesses some 52 000 km² of humid tropical land, about 44 per cent of its territory. The climate of this area is much more humid than that of the rest of the country; rainfall ranges from 2000 to 6000 mm annually (Jones, 1990, p. 62). In this section two colonization areas in the humid tropics of Nicaragua will be compared, one in the older agrarian frontier and one in the recent frontier. Two aspects will be covered:

- a comparison of the different stages of development of the two colonization areas;
- a comparison of production systems and farm development strategies in the two areas, aimed at finding entrances for the introduction of more sustainable agricultural practices.

It is relevant to compare stages of development of colonization areas because the main aspects relating to sustainability differ between areas at different stages of colonization. In the recent agrian frontier the important issue is to *prevent* rapid land degradation by introducing more sustainable and productive practices. The question is whether farmers can be expected to adopt such practices. Do they fit into the colonists' farming systems? Are there incentives inducing farmers to adopt them? Do farmers receive appropriate assistance? In the older colonization area farmers are often locked in into a low productivity trap; soil and pasture degradation have decreased productivity to levels at which maintaining subsistence production requires an increasing input of resources. The main issue is first to *restore* and then to *maintain* soil fertility and the carrying capacity of pastures by investing in land improvement. The question is whether farmers here have the capacity and the incentive to invest in soil and pasture reclamation.

Nueva Guinea, an example of the older agrarian frontier

The process of incorporation of the Nueva Guinea region into the national economy started in the 1930s with the extraction of rubber, raicilla

(*Cephaelis ipecacuana*) and hardwood (Ambrogui, 1995, p. 3). Colonization started in the 1960s when the first groups of peasants arrived.

Official colonization started in 1965 in the area of El Rama and Nueva Guinea with the Project Rigoberto Cabezas (PRICA) of the former Nicaraguan land reform institute, Instituto Agraria de Nicaragua, IAN. The stated objective of the project was to resettle peasants displaced from the western lowlands by the expansion of cotton production. The settlement of peasants, however, was also meant to create a workforce for large ranches in the colonization area, in particular for deforestation and sowing of pastures (Jones, 1990, p. 66). At the start of the official colonization project it was widely believed that most of the land in the agrarian frontier was still unoccupied. However, when the first inventories were made it became clear that the number of colonists already living in the Nueva Guinea area far exceeded the initial estimates. The population in the Nueva Guinea area increased rapidly in the 1970s and the first half of the 1980s; in 1994 there were 25 colonies in the Nueva Guinea area with about 17 100 families.

Nueva Guinea is an example of the older agrarian frontier; 80 per cent of the farmers in the area settled before 1980 (Table 5.1). Most of them originally came from the Pacific Coast and had been displaced by the expansion of commercial agriculture. But the largest part first settled in the Interior Region, especially in Boaco and Chontales, before they were forced out again, this time by the expansion of ranching in that region.

Many of the farmers questioned in the survey have lived in Nueva Guinea for 15 to 30 years. Despite land degradation, probably only a small proportion of colonists have sold their land in order to re-migrate to the new agrarian frontier. In the 1970s state assistance for official settlement, such as land titling, credit and extension services, kept re-migration low; in the 1980s the armed conflict reduced this option. This is not to say that the land tenure situation did not change. The colonists who stayed in the area were confronted with poor conditions for permanent cultivation of subsistence and cash crops, because of soil degradation and weed problems. According to the survey data part of the farmers sold the land they had acquired originally and bought new land in the area; others who stayed on the initial 50 *manzanas*[2] (mz) assigned to them by the land reform did some additional buying or selling of land, to enlarge or reduce their farm. In the course of time social differentiation of farm types increased and now five classes of producers can be distinguished (PRODES, 1992, p. 2):

Class 1 : the sharecropper. this type of producer owns less than 10 mz of land which moreover is not longer suitable for crop production due to depletion of soil fertility; land for food production is rented; livestock is of limited significance. The survival strategy of this producer often includes hiring out of labour;

Class 2 : the subsistence crop producer. although owning about 50 mz of land this type of producer is locked into subsistence production; his yields in food crops are low, implying that he must cultivate a relatively large area; the area in pastures is substantial, but on average he only keeps 6 head of cattle on 28 mz; the ratio of cattle to pasture is low (1:4.7). This type of producer has so far failed to escape the low productivity trap resulting from soil degradation and lack of capital accumulation;

Class 3 : the subsistence livestock – crop producer. the average farm size in this class suggests a small increase over time in area; production is still mainly for subsistence, but this type of producer aims at producing a small marketable surplus of food crops; the ratio of cattle to pastures is still low (1:2.8), on average 13 to 36 mz, but the sale of animals and dairy products is part of the survival strategy. The small base of market participation leaves little room for capital accumulation or for productivity increases;

Class 4 : the peasant livestock producer. with 39 head of cattle and an average farm size of 103 mz (1:2.6) this type of producer has more prospects of accumulating; because of the poor agro-ecological and socio-economic conditions of the area the accumulation so far has been of livestock, increasing the stock without giving attention to improvement of breed and pastures; crop production is mainly for subsistence, but if market opportunities arise this type of producer has the capacity to use them. A limited capacity for capital accumulation is used very narrowly in one direction, increasing the stock of cattle;

Class 5 : the farmer livestock producer. with 90 head of cattle and an average of 170 mz (1:1.9) this farmer is no longer a peasant type of producer but a farmer with an entrepreneurial strategy; 90 per cent of income comes from the sale of animals and dairy products; the main part of labour is hired, family labour is limited; this type of producer often transports or trades livestock; crops are mainly for consumption on the farm, and a considerable proportion of the grains is fed to pigs and poultry. While under-utilized pastures are available in the other farm types and renting of pastures is cheap, investment in land and land improvement is not a rational strategy. Accumulation is more diversified than in the case of the peasant livestock producer, but as far as agriculture is concerned marginal attention is paid to agro-ecological problems.

La Aurora, an example of the recent agrarian frontier

The hinterland of Bluefields in the Southern Autonomous Atlantic Region, the RAAS, is a lowland area characterized by dense rain forest, strips of swamps and mangroves, and deforested plots used for agriculture and live-stock. It is an area of dispersed homesteads and a few settlements built by the former Sandinist government to relocate rural people during the armed conflict (Vernooy, 1992, p. 25). There is almost no road infrastructure; most transport is by boat.

Presently the population of the hinterland of Bluefields amounts to about 7000 people. Migration from the Pacific and Interior regions started at the beginning of this century when foreign rubber, banana and lumber companies attracted both workers and peasants (ibid., 1992, p. 26). In the 1950s and 1960s when commercial agriculture developed in the Pacific zone and ranching in the Interior region, migration increased. The armed conflict in the second half of the 1980s restricted migration in the area to movements of peasants within the region, but since 1990 migration from the Interior has been increased again.

La Aurora is a recent (1990) settlement project for 150 families of Nicaraguan refugees who lived in Costa Rica at the end of the 1980s. The project is located at San Francisco, 75 km upstream the Kukra River. It is a so-called *polo de desarrollo*, a type of settlement project agreed upon in the 1990 peace negotiations between the government and the National Resistance Movement (RN). The government agreed to assign land to the demobilized, the repatriated and the refugees of the RN. At the request of the 150 refugee families which in 1990 settled on property made available to them by the church of San Francisco, the Nicaraguan land reform insti-tute INRA assigned 50 per cent of land to each family in the Caño Azul zone, two to three hour's walk from San Francisco. Most of them also bought or rented additional land at San Francisco for subsistence cropping.

Most of the colonists (69 per cent) in the La Aurora settlement come from the Atlantic Coast of Nicaragua (Table 5.1). Among them 25 per cent come from Nueva Guinea and El Rama reflecting the re-migration process. So far there is very little differentiation in farm type in the project; most farmers are of class 2, only a few of class 3, other classes have not yet developed.

Class 2 : the subsistence crop producer. 94 per cent of the colonists in La Aurora are classified in this class; the average peasant possesses 60 mz of land, of which only 20 per cent is used for crop and livestock production, the rest is still under forest or bush vegetation; on average there is only

Table 5.1   Comparison of Nueva Guinea and La Aurora settlements: general information[3]

Nueva Guinea	*La Aurora*
	1. PERIOD OF SETTLEMENT OF COLONISTS
Before 1980 (15–30 years)	After 1990 (< 5 years)
	2. ORIGIN OF THE COLONISTS (%)
Interior (66)	Atlantic Coast (69)
Pacific (12)	Interior (16)
Atlantic Coasts (9)	Pacific (4)
(No information: 17)	(No information: 12%)
	3. HOW DID THE COLONISTS GET ACCESS (%)?
Purchase (76)	Land reform (53)
Land Reform (20)	Purchase (31)
Inheritance (4)	Inheritance (16)
Additional purchase (20)	
	4. CLASSIFICATION OF FARMS

Advanced differentiation (class 1–5):	1992	1995	Little differentiation (class 2 and 3):	1995
1. Sharecropper.	34	–		–
2. Subsistence producer.	180	33	2. Subsistence producer.	48
3. Subs. livestock-crop prod.	68	15	3. Subs. livestock-crop prod.	3
4. Peasant livestock producer.	41	4		–
5. Farmer livestock producer.	16	2		–

Table 5.1 Continued

5. SIZE OF FARMS (BY CLASS)

	Nueva Guinea					La Aurora				
Class	Total (Mz)	Crop	Pastures	Forest	Livestock (head)	Total (Mz)	Crop	Pastures	Forest	Livestock (head)
1.	5.0	2.4	2.4	0.2	3	–	–	–	–	–
2.	50.2	8.7	28.0	13.5	6	60.6	5.8	5.6	49.2	1
3.	57.3	9.3	36.3	11.7	13	66.7	20.7	14.2	12.6	22
4.	103.0	12.7	75.4	14.9	39	–	–	–	–	–
5.	168.6	8.2	151.7	8.7	90	–	–	–	–	–

one animal per farm, in this class to accumulate livestock has not yet started; there is a small marketable surplus of beans and maize, but marketing is difficult because of lack of transportation facilities;

Class 3 : the subsistence livestock and crop producer. only 6 per cent of the colonists in La Aurora are classified as class 3; this type of producer has about 66 mz of land and owns 22 head of cattle on average. Most of the cattle initially distributed among all the colonist-refugees in the project were bought by the farmers in this class who started the process of accumulating livestock.

After clearing of the forest land degradation in Nueva Guinea has advanced rapidly and is widespread in the area. As re-migration was low in the 1970s and 1980s the pressure on the land built up. Sustainable development now requires soil and pasture reclamation through practices such as cover crops, home gardens, alley cropping, pasture renovation and silvopastoral systems. These practices imply investment in restoring soil fertility (nitrogen fixing, recycling of nutrients), improving weed control and tree planting. The benefits of these investments are not immediate. Farmers are only willing to make such investments when land for crop production or pasture land becomes scarcer because of favourable prospects for production and marketing. A case in point is the recent demand for *quequisque*, a tuberous crop in great demand from Costa Rica. Producers are increasing the use of fertilizers in this crop and are more interested in crop rotation; there are some credit facilities for this crop.

As can be observed from the classification of producers, many of the farmers have no capacity for capital accumulation and to reverse depletion of soil fertility and weed infestations. Those farmers that have some capacity to invest, invest in cattle rather than in land improvement.

In La Aurora land degradation so far is limited; soil fertility and weed problems are still manageable. Although subsistence production requires less resources than in Nueva Guinea, market production is even more constrained by market outlets; services of all types are lacking. Diversification of subsistence production, particularly with tree crops, can simultaneously improve consumption and sustainability; but colonists also have an increasing need for cash income. If production for the market cannot be developed based on permanent crops, or by giving value to the forest and forest products, the colonists in this area will also end up with extensive livestock production as their main mechanism for participation in the market. But if sustainable forms of crop and forest production develop, the area of degraded land – which can be used only for extensive ranching – will stay small. Options in this recent agrarian frontier are still open.

4   PRODUCTION SYSTEMS AND FARMERS' DEVELOPMENT STRATEGIES

In this section production systems, farmers' development strategies, adoption of alternative practices and perceptions on re-migration are discussed. Can farmers be expected to adopt more sustainable practices? Do these practices fit into the existing farming systems? Are they compatible with farmers' perceptions of farm development and re-migration?

Production systems

Food grains predominate in the production systems of both areas (Table 5.2). In Nueva Guinea the majority of the producers, including the livestock producers, cultivate maize, beans and rice for home consumption and sale of a marketable surplus. On the larger farms, classes 4 and 5, the yields of maize and beans are higher because these producers are able to select the better part of their farms for crop production (PRODES, 1992, p. 33) and to maintain some form of rotation. In maize the class 4 and 5 farms have a larger volume of production, but proportionally they sell less on the market because more is consumed by their pigs and poultry. In the case of beans, production and sales both increase with farm size. Rice is consumed on the farm and sold in the market, but is a minor commercial product. As indicated, during the last five years a market has developed for *quequisque*; all farm types cultivate this commercial crop on a modest scale, the area increases with farm size.

Keeping livestock is an important activity for all farmers; animal production represents less than a third of income for class 1, compared with half the total income for classes 2 and 3, and 80 per cent for class 4 and 90 per cent for class 5. The smallest producers, the sharecroppers mainly keep pigs and poultry; the class 2 peasants sell pigs and calves; class 3 adds dairy products and cows to these sales and classes 4 and 5 multiply the scale of animal production.

In La Aurora the main crops are maize and beans; class 2 farmers sell about a third of the maize and half of the beans, which are the same proportions as found for the class 2 farmers in Nueva Guinea. According to the information obtained in the survey, yields in La Aurora still are about double those in Nueva Guinea; consequently a much smaller area per farm is planted with maize and beans. Hardly any rice is grown and other cash crops are rare. Animal production is the distinguishing difference between class 2 and 3 producers; class 3 farmers sell milk in San Francisco, but

Table 5.2 Production systems and grade of commercialization

Nueva Guinea	*La Aurora*

1. PRODUCTION SYSTEM

Nueva Guinea	La Aurora
Foodgrains: maize, beans, rice.	Foodgrains: maize, beans.
Cashcrops: beans, tuberous crop (*quequisque*).	Cashcrops: beans
Livestock for subsistence (class 1 & 2)	Livestock: mainly for subsistence
Livestock & livestock products (class 4 & 5).	

area (mz)	yield (qq')	production (qq)
7.0	5.3	37.1
3.9	5.2	20.3
1.3	4.5	5.9
1.5	14.3	21.5

crop	area (mz)	yield (qq)	production (qq)
maize	2.9	14.5	42.1
beans	2.5	10.7	26.8
rice	–	–	–
quequisque	–	–	–

Fertilization (58%): chemical, organic, green manure.
Herbicides (63%): low cost, labour saving, quick result.

Fertilization (14%).
Herbicides (35%).

2. GRADE OF COMMERCIALISATION

Consumed on farm (%)	Sold (%)
67	33
49	51
45	55
5	95

Class	Consumed on farm (%)	Sold (%)
Class 2 maize	63	37
Class 2 beans	49	51
All classes arroz	–	–
All classes quequisque	–	–

outlets for dairy products such as cheese and curds are limited (Lopez & Zeledón, 1995, p. 13).

Farmers in the area have made some adaptations to the environmental conditions they face; they do very little ploughing with oxen or tractors in order not to disturb the fragile and shallow soils. Fertilization is much more frequent in Nueva Guinea, with 58 per cent of the farmers using chemical or organic fertilizers, or green manure. In La Aurora where depletion of soil fertility is less general, only 14 per cent of the peasants use some form of fertilization. The same situation is found in the case of herbicides; in Nueva Guinea, where weed infestation is a severe problem, 63 per cent of the farmers apply herbicides. These are used because of their relatively low cost, their labour saving character and quick results. Only 35 per cent of the colonists in La Aurora use herbicides; weeds are not yet a severe problem.

Farm development strategies

As far as priorities for development of the farm are concerned, there are no significant differences between the colonists in Nueva Guinea and La Aurora. For one third of the farmers the development of livestock production is the first priority. Ranching is seen as a viable activity because it is affected less by depletion of soil fertility than is the case with crop production. A lower carrying capacity of pastures can be compensated for by running the cattle on unused land, available on most of the farms. By contrast, in crop production the decreasing yields translate into higher unit production costs. Livestock and livestock products also have a more secure market than crops; moreover, livestock products generate a more regular income. Livestock is also considered an appropriate mechanism for capital accumulation. In the La Aurora settlement the high priority assigned to livestock is remarkable, because only 3 per cent of the farmers have started to build up a small herd; many more perceive animal production as a viable road to integration in the market and to capital accumulation.

In the Nueva Guinea area the percentage of farmers giving preference to an increase in the number of animals exceeds the proportion of producers giving priority to the improvement of pastures – including tree planting – and to the increase of area. In La Aurora, as there is only a small number of livestock producers, the strategy for development of animal production is not yet sufficiently defined; but the few farmers who have developed livestock activities already had a smaller percentage of land under trees.

One quarter of the colonists surveyed in both areas indicate subsistence crops as their first priority, which shows the importance of farm consumption by people as well as by animals. A much lower proportion of farmers indicate cash crops as their first priority, far behind animal production as a commercial activity. Over 80 per cent of the colonists in La Aurora emphasize the need to increase the area under crops. In Nueva Guinea, in contrast, farmers give priority to augmenting yields (49 per cent) and less to increasing the area under crops. This again reflects the critical situation of depletion of soil fertility in this zone. In both areas planting of perennial crops has a rather low priority for farmers, although a substantial proportion have planted a variety of trees in the farmyard, mainly for home consumption (citrus, mangos, bananas and *pejibaye*).

Sustainable production practices

The adoption of more sustainable production practices is a crucial aspect, not only to prevent but also to redress land degradation in the colonization areas in the humid tropics. Farmers in both areas have some knowledge of these practices; some have experimented with them. In the Nueva Guinea area, governmental and non-governmental organizations are encouraging farmers to incorporate these practices into their production system. There are two experimental farms in Nueva Guinea, and a number of farmers run long-term experiments on their own farms. The extension method widely accepted in the area by governmental and non-governmental agencies is based on the farm classification discussed above and on a farming systems approach. New practices are only introduced to farmers if they fit into the specific production system. A farmer-to-farmer model is used; around the farms directly visited in the extension programme (the Reference Farms – RF), groups of farms have been organized (No-Reference Farms – NRF) to work together with the RF on the selection, the experimentation and the evaluation of the innovative practices.

The fertilizer bean–maize rotation

In the Nueva Guinea area using of fertilizer beans to restore fertility in degraded soils does not produce positive results until a number of cycles. But more immediate benefits have been obtained by combining a fertilizer bean rotation with improved practices in maize production, such as seed selection, planting distances, and varieties (Sanclemente *et al.*, 1994, p. 57).

As can be observed in Table 5.3, 61 per cent of the farmers in Nueva Guinea have firsthand knowledge of this combined practice, and 59 per cent of these farmers indicate that they have experience with it. A high proportion of all farmers indicate that they are planning to apply this practice in the future. In Table 5.4 the same information is given for RF and NRF separately. More of the RF farmers know of the fertilizer bean–maize rotation and a high percentage of them have tried it. The corresponding figures for the NRF are much lower, but proportion of farmers planning to use this practice in the future is the same. In La Aurora the percentage of farmers that know the fertilizer bean–maize rotation is not

Table 5.3   Adoption of more sustainable practices by colonists (percentage yes)

Nueva Guinea		La Aurora
	1. Velvet bean in rotation with maize and beans (%)	
61	farmers have first hand knowledge	61
59	experience with this practice	59
83	farmers plan to use this practice	83
	2. Alley cropping (%)	
46	farmers know this practice	35
20	experience with practice	2
60	positive experience	–
44	farmers plan to use this practice	29
	3. Silvopastoril systems; reasons for planting trees (%)	
80	shadow for animals	61
44	soil fertility	26
44	live fences	47
22	cattle eat leaves and twigs	20
76	farmers plan to plant trees	75
	4. Home gardens (%)	
–	farmers have home gardens	16
	5. Who is providing services (%)?	
44	the state	18
7	the bank	13
22	non-governmental organization	26
4	neighbours, other persons	4
23	no services received	39
	6. Is there a farm that serves as a model for the colonist (%)?	
80	yes	48

Table 5.4   Comparison of reference and non-reference farms in Nueva Guinea, 1995 (in%)

Issue	All farms	Reference farms	Non-reference farms
1. Input use			
Use of chemical fertilizers	26	41	19
Use of organic fertilizers	13	24	8
Green manure	44	65	35
Use of herbicides	63	53	68
2. Priorities for farm development			
Livestock production	35	41	32
Subsistence crops	24	24	24
Cash crops	22	29	19
Simultaneously	15	6	19
No information	4	–	6
3. Priorities in crop production			
Increase area subsistence crops	20	18	22
Increase yields of subsistence crops	50	64	43
Increase area perennial crops	11	6	14
Increase area annual crops	4	–	5
No information	15	12	16
4. Priorities in livestock production			
Increase stock	32	23	35
Increase area under pasture	9	12	8
Improve pastures	9	6	11
Tree planting on pastures	11	18	8
No information	39	41	38
5. Adoption of velvet bean in rotation with maize and beans			
Farmers have first hand knowledge?	61	65	59
Do farmers that know this practice Perceive results as positive?	59	82	49
Do farmers plan to use this practice?	83	82	84
6. Adoption of alley cropping			
Do farmers know this practice?	46	65	38
Do farmers that know this practice Perceive results as positive?	60	64	57
Do farmers plan to use this practice?	44	59	38
7. Adoption of silvopastoril systems: reasons for planting trees:			
Shadow for animals	80	82	78
Fertility	44	47	43
Live fences	44	71	32
Cattle eat leaves and twigs	22	35	16
Do farmers plan to plant trees?	76	76	76

Table 5.4   Continued

Issue	All farms	Reference farms	Non-reference farms
	8. *Do the farmers receive services from:*		
The state	44	65	35
The bank	8	12	5
A non-governmental organization	22	18	24
Neighbours, other persons	4	–	6
No services received	22	5	30
	6. *Does the farmer perceive possibilities for himself for further migration to the Agrarian Frontier?*		
Farmers considering migration	20	6	27
Not considering migration	72	85	65
Doesn't know	8	6	8

much lower than in Nueva Guinea; most of these farmers have observed this system in Costa Rica. However, no one in La Aurora has tried it.

Alley cropping

Experiments with this practice were run for the last ten years in Finca Esperancita, one of the two experimental farms in the Nueva Guinea area, and it was then introduced to the RF of Finca Esperancita. Less than half of the farmers surveyed said they knew about alley cropping; of those, only 60 per cent think that this system has positive results. Only 20 per cent of the farmers have tried this practice; and 44 per cent of all farmers plan to try it in the future. There is quite a difference in knowledge, experience and planned use with respect to this practice between the RF and the NRF. In La Aurora the knowledge of alley cropping is limited and was mainly obtained in Costa Rica; experience is nearly nil; nevertheless, 29 per cent of the farmers are thinking about using this practice in future.

Silvopastoral systems

There are different reasons for tree-pasture combinations. The one mentioned most frequently by farmers in Nueva Guinea was shade for animals; live fences and soil fertility were mentioned in less than half of the cases and browsing of leaves and twigs was indicated by only a fifth of the farmers. There was little difference between RF and NRF.

Nevertheless 76 per cent of all farmers, RF and NRF, plan to plant trees in pasture-tree combinations. In La Aurora livestock keeping is still in its infacy but 75 per cent of the farmers plan to plant trees in pastures.

Home gardens

In La Aurora only 16 per cent of farmers have what can be considered as the beginning of a home garden. No information as this is available for Nueva Guinea. PRODES is including this mixed tree system in its recommendations to RF.

As can be observed from these findings, systems like the fertilizer bean – maize rotation and the planting of trees for shade in pastures are well known practices. They are relatively easy to transfer, partly because farmers perceive the results as beneficial. There is not much difference between RF and NRF in this respect. But other practices, such as alley cropping or silvopastoral systems, are less known and their possible benefits less understood. RF farmers have a lead in these practices, probably partly because of the extension services they received as RF. As these practices, are more complicated and require more labour inputs, whereas their benefits are less visible and less easy understood, extension services must stimulate all farmers involved to experiment on their own farms.

80 per cent of the Nueva Guinea farmers mentioned one or more farms that serve as a model for the development of their own farm. Apart from the two experimental farms of the area, these model farms include farms on which the farmers experiment on their own, and the RF. The answers clearly reveal the relevance of the farmer-to-farmer extension model. In La Aurora only 48 per cent of the colonists have a farm which serves as an example for the development of their own farm.

Further migration to the new agrarian frontier

Table 5.4 also touches on the question of the pattern of re-migration to the agrarian frontier. Only one fifth of the farmers in Nueva Guinea perceive the possibility of further migration to the new agrarian frontier; although many more see opportunities for their children. In the case of re-migration, most of these farmers think land in the new frontier must be bought; squatting and land reform are no longer the most important ways to gain access to land. In La Aurora the proportion of colonists possibly considering re-migration to the new frontier is higher, as is the percentage of farmers still counting on the possibility of squatting. The colonists' perception of the possibilities for further migration to the new frontier is an important factor in shaping their attitude on more sustainable practices and land use systems in their actual location.

5 CONCLUSIONS

Reducing the flow of migrants to the agrarian frontier is an important mechanism to protect the rain forest, but economic policies in the 1990s have not resulted in a decrease of colonization. The alternative approach to limit the negative impact of colonization is to introduce more sustainable practices and land use systems in the colonization areas.

In colonization areas farmers use extractive production systems. Although the availability of land decreases and follow periods are too short to assure a natural recovery of soil fertility, farmers still go to the nutrients, instead of nutrients to the farmers.

Land is becoming scarce *for the individual farmer* as he no longer has access to free land and lacks the means to buy it. He is mining the soils. As yields fall, costs of subsistence and cash crops rise, reducing the viability of crop production. The remaining option for degraded land is use as pasture, but as crops become less profitable producers lack the means to invest in cattle.

Some of the colonists get locked into a low equilibrium trap. Without cattle the use of their land is limited. Land can be rented out (cheaply) to cattle owners; at the same time land must be rented in (at a higher price) for subsistence production. This situation is most beneficial for farmers with cattle or access to credit for livestock. Social differentiation intensifies.

As the limits of the agrarian frontier are about to be reached, land is also becoming scarce *for the community*. Accordingly, extractive production systems have to be replaced by more sustainable ones, in which nutrients extracted by production are – at least largely – replaced.

The main question for policy design is how to promote the transition to more sustainable practices and land use systems. The findings of this study indicate several areas where incentives can be developed.

In the *older colonization* areas

- credit facilities for farmers with a small number of animals or without cattle, to enable producers to increase the use of land at their disposal. The amount of land available for rent decreases and livestock production increases, pasture land will become relatively scarce and incentives to invest in land and pasture reclamation will increase. The survey indicates that livestock development is the first priority of colonists;

● yield increases in subsistence crops are important; decreasing yields have resulted in greater efforts being required for home production of food grains. This implies a higher time preference for subsistence production and less labour available for improvements on the farm. Yields of subsistence crops can be increased with rather simple practices: maize – fertilizer bean rotations; alley cropping and home gardens. As the survey indicates, knowledge and practical experience play an important role in the introduction of these practices. Extension services in this field can make a crucial contribution;

● as the *quequisque* case indicates, improved marketing opportunities create incentives to invest in more sustainable practices. Improving marketing organization of the farmers and marketing services (financing, processing, storing) – in particular for selected tree crops – is instrumental in facilitating the transition to more sustainable practices and production systems;

● production of wood and forest products can become an alternative source of income for colonists, as the area of primary forest is decreasing rapidly. So far farmers in the colonization area lack knowledge and experience in this.

In the *recent agrarian frontier*:

● options to *prevent* land degradation still exist: early attention to the introduction of sustainable practices and land use systems, especially in subsistence and market crops, and giving value to forest production can prevent a fall in productivity and will make extensive cattle ranching a less attractive option.

Notes

1. The assistance of Alfons Franssen in the elaboration of the survey data is gratefully acknowledged.
2. Manzana (mz): 0.7 hectare.
3. The data in Tables 5.1 to 5.4 are based on a survey made in 1995 in the Nueva Guinea and Bluefields (La Aurora) areas. The Nueva Guinea survey included 54 farmers, the La Aurora survey 51 farmers. The information in Table 5.1 is partly supplemented with data from a published survey from PRODES (a survey under 391 farmers, 1992). In Table 5.4, only for the Nueva Guinea data, the distinction is made between farmers directly served by governmental and non-governmental extension agencies on aspects of

88   *Production Systems in Nicaragua*

sustainable production practices (17 Reference Farmers) and those indirectly served (in groups) or non-served (37 Non-Reference Farmers).
4. qq (quintal): 45.45 kilogramme (100 pounds).

Bibliography

Ambrogui, R. (1995) *Caracterización de los sistemas de produccción en Trópico Húmedo. Colonistas en el municipio de Nueva Guinea*, (Managua: ESECA)

Bakker, M.L. (1993) *Colonization and Land Use in the Humid Tropics of Latin America*, BOS, Wageningen.

Howard-Borjas, Patricia (1992) *Cattle and Crisis: The Genesis of Unsustainable Development in Central America. Case studies of Honduras and Nicaragua* (Rome: FAO).

International Fund for Agricultural Development – IFAD (1992) *Soil and Water Conservation in Sub-Sahara Africa. Towards sustainable production by the rural poor*, A report prepared by CDCS, Free University, Amsterdam (Rome: IFAD).

Jones, Jeffrey R. (1990) *Colonization and Environment. Land Settlement Projects in Central America* (Tokyo: The United Nations University Press).

Kaimowitz, David (1995) *Livestock and Deforestation in Central America in the 1980s and 1990s: A Policy Perspective*, Forthcoming.

López Jiménez, Mario & Ramón Zeledón (1995) *Caracterización de los sistemas de producción en Tropico Húmedo. El polo de desarrollo de La Aurora-San Francisco Kukra River* (Managua: ESECA).

Maldidier, Christobal *et al.* (1993) *Tendencias Actuales de la Fronter Agrícola en Nicaragua* (Managua, NITLAPAN-UCA).

Pasos, Rubén, coordinador (1994) *El Ultimo Despale ... La Frontera Agrícola Centroamericana*, FUNDES-CA (San José: Gamier Relaciones Públicas, S.A.).

Programa de Desarrollo Rural en la Zona de Nueva Guinea – PRODES (1992) *PRODES Diagnóstico Socio-Económico* (Nueva Guinea: PRODES).

Ruthenberg, Hans (1976) *Farming Systems in the Tropics*, second edition (Oxford: Oxford University Press).

Sanclemente, Oscar H., Salas R. Ligia María & Bruno Molijn (1994) *Resultados de los Ensayos de la Producción de Maíz en rotación con Frijol Abono (ciclos 1992/1993 y 1993/1994)* (Nueva Guinea: PRODES).

Utting, Peter (1991) *The Social Origin and Impact of Deforestation in Central America*. Discussion Paper 24, UNRISD, Geneva.

Vernooy, Ronnie (1992) *Starting All Over Again. Making and Remaking a Living on the Atlantic Coast of Nicaragua*, Doctoral Thesis, Wageningen Agricultural University,

Williams, R.G. (1986) *Export Agriculture and the Crisis in Central America* (Chapel Hill: University of North Carolina Press).

6 Food Insecurity as a Sustainability Issue: Lessons from Honduran Maize Farming*

Hazel Johnson

1 INTRODUCTION

This chapter is concerned with the vulnerability of small producers of food staples in Honduras to food insecurity in the 1980s. By vulnerability to food insecurity, I mean the constant and endemic threats to such farmers' being able to reproduce their production of and access to adequate food staples. The context is one of inequality in access to land and other resources for production, uneven commercialization and limited opportunities of alternative employment for small farmers. During the 1980s, the Honduran government and other institutions attempted to improve the productive capacities of individual and collectively-organized small farmers by credit and technical assistance packages to encourage diversification, as well as increase national output of marketed food staples to reduce the need for food imports. However, such policies did not always take into account structural obstacles to change. The question remains as to whether such farmers (and their children) can sustain and improve their productive capacities in the longer-term.

The concept of sustainability being used in this chapter draws on broad definitions. Although I am particularly concerned with the socio-economic processes that affect small farmers' capacities to sustain their production of and access to food staples, definitions of sustainability from the environmental literature are relevant. Thus the Bruntlandt Report's definition,

---

* A revised version of this article appeared as H. Johnson (1996) 'Vulnerability to food insecurity among Honduran maize farmers: challenge for the 1990s', *Journal of International Development*, Vol. 8, No. 5 pp. 667–82.

'Sustainable development is development that meets the needs of the present without compromising the ability of future generations to meet their own needs' (WCED, 1987), is pertinent in the context of land fragmentation and degradation among small farmers in Honduras which provides an uncertain future in farming for their children. The vision laid out in Latin American response, Our Own Agenda, includes a political manifesto for participatory democracy (Gabaldón, 1992). This is particularly important when the abilities of small farmers to influence change are structurally constrained. Thus sustainability is about human capacity building to meet fundamental needs and improve the quality of life (Munslow *et al.*, 1995).

The analysis in this chapter focuses on the sustainability of small farmer production of a particular food staple, maize. Maize is widely produced by small farmers for consumption and the market in Honduras, as well as by collectively-organized peasant groups, cooperatives and large commercially-oriented landowners. Its role in providing food and cash income is a key aspect for the daily livelihoods of many rural people. In the 1980s, Honduran policy-makers attempted to increase national output of maize for the market to meet the needs of urban and rural waged consumers and reduce imports. Policies were thus directed to the commoditization of production processes and output as well as to the food and income needs of the producers. Although the production of marketable surpluses has been a key issue, maize is a low value crop commercially which requires considerable labour time. Thus small farmers helping to provide maize for waged consumption in towns and rural areas are often vulnerable to food insecurity because of low cash incomes and inadequate self-provisioning.

Focusing on a single crop is not a means of assessing the overall capacities of farming households to sustain their production and access to food. Small maize farmers in Honduras will also try to produce the other main staple, beans, as well as some vegetable and fruit crops, keeping small animals, engaging in different forms of wage work and self-employment, and they may receive remittances from children or other relatives. However, focusing on maize enables me to analyse (i) the difficulties of sustaining production of a low value food staple for consumption as well as the market in an increasingly commoditized economy; (ii) the social relations which both restrict and as well support the productive capacities of small producers, and (iii) the resulting difficulties of diversifying and improving the such farmers' productive capacities more generally.

The chapter is based on research into the vulnerability of semi-proletarian and petty commodity maize farmers in the mid to late 1980s. As well as extensive analysis of existing data, field research was carried out among producers (individual and collectively organized) in two villages in the department of El Paraíso, among collectively organized producers in Santa Bárbara, and among traders in the towns of Danlí and San Pedro Sula. The research has shown the difficulties experienced by semi-proletarian farmers and petty commodity producers in reproducing maize production on a sustained basis without different forms of personalized and institutional indebtedness (Johnson, 1995). Moreover, such farmers were often unable to keep adequate quantities of maize from their output for their own consumption, and frequently earned negative net incomes from maize.

The chapter first addresses some conceptual issues in explaining the vulnerability of many maize farmers. It then looks at three key policy areas for rural livelihoods in Honduras in the 1980s: land, credit and output markets. Some lessons for 1990s policy are drawn.

2   ANALYSING VULNERABILITY AND RURAL FOOD INSECURITY

Food insecurity is seen as arising from difficulties in people's capacities to reproduce the production of sufficient food or other means of access to it. In particular, it involves analysing why and how people's endowments and entitlements can be threatened (Sen, 1981). My concern is to look at the threats posed to endowments and entitlements by relations of exchange. Exchange is the main means of obtaining resources for production as well as realizing its benefits. Thus analysing exchange relations among maize farmers reveals how access to resources for production and access to the benefits of production may be reinforced or made vulnerable.

Since the 1970s, there is a considerable literature from and about Latin America on changing agrarian structures, the development of markets and agricultural and food policy (for example, Bartra, 1976; Oswald, 1979; de Janvry, 1981; Austin and Esteva, 1987; Brockett, 1990; Hewitt de Alcántara, 1992). In addition to the richness of this literature, illuminating writing on exchange relations under capitalist development has come from India (for example, Bhaduri, 1983; Bharadwaj, 1985), as well as from debates on commoditization and changing social relations (for example, Long *et al.*, 1986; Scott, 1986), and on the social and political nature of markets (for example, Hewitt de Alcántara, 1993; Mackintosh, 1990; White 1993). Among other important contributions, this literature has shown that (i) commodity markets may not develop at the same pace, (ii) exchanges can involve non-price and non-market processes such as tied transactions, personalized debt relations, and coercion usually based on the social positions of actors and relations of dominance, and (iii) approaches which analyse markets as institutional processes of buying and selling can help analyse the effects of market exchanges on people's lives and livelihoods.

I have been particularly concerned to analyse the effects of commoditization on maize farmers' capacities to sustain production and to distinguish the appearances and realities of commoditization.[1] Because commoditization is an uneven process with differential effects, farmers may seek other means of obtaining access to resources or exchanging goods and services. Such processes may be pursued as active strategies, but they may also lock farmers into relations which affect their productive capacities.

The example of access to land in Honduras provides an illustration. Although land markets existed in the 1980s, access to land for many small farmers was by loaning, or exchanging land for labour (as well as by occupying nationally- or municipally-owned land). Landowners and tenants were aware of the rental and market value of land, but arrangements over access could be defined by other criteria – for example, whether tenants could provide wage labour to landowners at harvest or whether the landowners were prepared to loan agricultural inputs to tenants.

These arrangements have a non-commoditized appearance but in practice they depend on the existence of commodity markets, especially those for credit (on the basis of which landowners loan inputs to tenants) and labour (in terms of setting wage rates). Such types of exchange are based on personalized relations. On one hand, they are not non-commoditized since they depend on commoditized relations and markets for their existence. On the other, they are not impersonal transactions which take place within the markets for land and labour: they depend on personal networks and relations between individuals and families which may also involve elements of reciprocity, however implicit and unequal.

It is important to distinguish between appearance and reality in other types of apparently non-commoditized relations, such as the use of family labour in production. For example, Brass (1986) has shown how surpluses are accumulated *within* petty commodity production in Peru through the unequal rights and obligations of kin and 'fictive kin'.[2] As well as being forms of reciprocity, kin and fictive kin relations can be used to extract surpluses, gain access to labour and keep wages down because of the supposed familial and reciprocal ties. In the process, kin and fictive kin relations themselves become commoditized.

Processes of exchange and their effects on the capacities of maize farmers to produce and sustain access to maize are embedded in social divisions and hierarchies, particularly class relations, and associated power and control. First, agrarian class formation and associated social and economic change can pose threats to livelihoods through displacement and proletarianization (see, for example, Bartra, 1976; de Janvry, 1981; Scott, 1986). Second, analysis of the social context and content of exchange relations indicates the importance of power. With respect to trade relations, Bardhan states: 'The concept of power goes beyond the outcome of a given exchange and points to the fact that power may be centrally involved in causing the existing pattern (and defining the existing parameters) of trade in the first place' (1991, p. 267).

In Honduran maize farming I distinguish three classes of farmer: commercial, petty commodity producers and semi-proletarian.[3] Commercial farmers predominantly use wage labour in production while semi-proletarian farmers cannot survive without selling labour. Exchanges between these two types of farmer are part of their production strategies. Petty commodity producers occupy an intermediate position – they hire some labour, but they do not sell labour; they may engage in simple reproduction or may be able to expand their production. These classes are also based on social hierarchies involving relations of power and patronage over access to land, and differentiated control over labour, technology and output. Actions of the state which intervene in these social hierarchies may reinforce or undermine the relative power which people have.

Class relations, power and the processes of exchange involve conscious action as well as structural constraints. For example, relations which may lock semi-proletarian farmers into subordinate exchange relations with landowners may both help to provide security as well as limit farmers' capacities to increase output or consumption of maize. Equally, joint action by semi-proletarian or landless producers can contest property rights and modes of access to land. Interventions by the state may also reduce or serve to reinforce vulnerability to rural food insecurity. These different types of action indicate the importance of understanding action and policy formulation as processes involving contested terrains, different interests and points of leverage (see Drèze and Sen, 1989; Wuyts *et al.*, 1992).

In sum, analysing exchange relations entered into by farmers to secure their means of production and exert some control over the distribution of output can help to explain rural food insecurity. Power and social hierarchies play an important role in these exchanges, as do forms of reciprocity. The exchanges can serve both to undermine or enhance farmers' capacities to continue in production, as can interventions by the state and other institutions in such a context of unequal social relations. Analysing the production and exchange relations of a staple which is a source of livelihood and food for most farmers can also help explain wider social relations of rural poverty and stagnation. In addition, it can illuminate some of the difficulties faced in Honduras of trying to raise national output of staple foods and reduce vulnerability to food insecurity in the countryside.

3　POLICIES AND PRACTICES IN MAIZE FARMING

Increasing awareness of food insecurity came on the policy agenda in Honduras in the 1980s for several reasons: evidence of malnutrition from surveys at the end of the 1970s; studies of urban and rural poverty; continuing landlessness in rural areas and low levels of employment and income; the growing need for food imports and aid; pressures for government action by peasant organizations. The 1970s had been a period of military-led reform in the countryside, including a major land reform act, which had been preceded by drives for change and modernization during the 1950s and 1960s, as well as a growth in peasant organizations and protests around lack of access to land and means of livelihood. State intervention in the economy had boosted the public sector, establishing institutions for directing and regulating finance, investment and marketing in agriculture, industry and services. Although the lives of some rural people were improved by these changes, rural poverty remained a widespread and chronic problem. As Honduras moved from military to civilian government in the 1980s, it also moved gradually from an era of state intervention in the economy to a period of increasing liberalization, deregulation, emphasis on private property and individual initiative. These changes were also to affect policies directed to rural livelihoods.

Land and labour relations

As in many Latin American countries, historically there has been a high level of inequality in land distribution. Estimates of the numbers of landless in the countryside range from 30–55 per cent of the economically active population (SIECA, 1984) or even higher, and there has been extensive migration between rural areas and between the countryside and towns. Historically, national or public lands have been an important source of use rights for poor farmers as well as for those who have appropriated large tracts of this land for cattle. However, many farmers do not have secure access to land: for example, expulsion of small farmers has also accompanied the expansion of pasture, and many without land rent or borrow from it other farmers.

Access to land has been a key area of policy and public action. Honduran legislation on land has experienced several metamorphoses, from periods of land reform in the 1960s and 1970s, during which a limited land distribution (and more limited redistribution) took place, to a process of titling national or public lands in the 1980s, to the reinforcement of privatization of tenure and land markets in the early 1990s. The changes in these policies paralleled changing perspectives on social and economic policy reform: from reformist state interventionism in the 1970s to the growing predominance of neo-liberal policies in the 1980s. The period of land reform attempted to create types of farm based on collective as well as individual property and presupposed state support for infrastructure, credit and technical assistance. More recent moves on land titling,[4] privatization of land tenure and the reinforcement of land markets have centred on creating and reinforcing individualized farming.

Continuing issues for land policy are how to provide land for the numerous landless in Honduras, and how to improve the land allocation and security of access for those small farmers who rent or borrow small plots. The latter in particular requires an appreciation of the types of exchange that such farmers enter into and how it secures or constrains their production of and access to food staples.

A small case study of six semi-proletarian maize farmers interviewed in two villages in El Paraíso illustrates this point. Field data show that these farmers' access to land depended on reciprocal land/labour relations embedded in a combination of commoditized and personalized exchanges. Furthermore, access to land in this way enabled such farmers to produce maize but limited their capacities to provide adequate maize for consumption as well as an obtain a positive net income from it.

Access to land for those without inherited plots, or national or public land, was usually by renting from commercial maize farmers. Commercial maize farmers were usually large land-owners who combined cattle with maize (and possibly beans and other crop) production. There were different arrangements for renting land, often expressed in the phrases '*me presta una manzana*', '*me da una manzana*' or '*me alquila una manzana*'.[5] They give a flavour of the personalized nature of the agreement. Sometimes rent was charged, and sometimes not; rent was usually in cash but may also have been in kind. There were other aspects to the exchange: tenants often provided labour to landowners. This was paid labour at local wage rates and may have varied in time and activities.[6] Furthermore, landowners (who received institutional credit) acted as a source of input loans which were repaid at harvest.

Thus, the exchange networks for these farmers showed a relationship between access to land and provision of labour. In some cases this exchange was reinforced by loans of agricultural inputs (Table 6.1). None of the six had access to more than about two Has of land of (one only had 1 manzana or. 7 Ha). Most rented or borrowed all their land and none rented or borrowed less than half. One of the six obtained land from his brother for whom he also did wage work. This farmer wanted to break this relation and become financially independent. However, his proposed solution – to obtain institutional credit – was problematic because he had no or little land of his own.

The small amounts of land available were barely enough for the food requirements of a household (which might average 6 people, although in one case it was 17). After debt repayments, only two of these farmers had positive net cash incomes from maize production while four of the six could not keep enough maize for household consumption needs (including animals and seed). However, their wage work and their ongoing access to plots loaned or rented by the landowner gave them means to continue

Table 6.1   Exchange networks of six semi-proletarian farmers interviewed in two villages of El Paraíso, primera[7] 1986–7 [8]

Farmer	Land	Wage work	Oxen hire	Tractor hire	Mechd. shelling	Weed-killer	Seed	Urea
AM	L1*	L1	L1	L1	L1	Other	L1	L1
RG	L2	L2	L2	L4	L4	L2	None	L2
JAC	L4	L4	None	None	L4	Other	None	L4
JC**	L2	L2	L3	L3	None	None	Other	None
JS	L1	L1	L1	None	L1	Other	Other	Other
RZ	Brother	Brother	Brother	L6	None	Other	Other	Other

L = Landowner.
*   AM had just started renting land from his landowner for whom he already worked.
** JC also financed production from pre-harvest sales.

producing maize as well as to buy it for household consumption. The problem is that they were always indebted to the landowner, were vulnerable to the personalized relations breaking down, often had to employ temporary wage labour on their own plots if they were working elsewhere, and often had to sell maize they needed for consumption to meet cash debts. Such deficit farmers also made maize purchases at higher prices than those received for their own output.

Other maize farmers with landless and semi-proletarian histories, who had occupied land as collectively-organized groups, were able to break with such land/labour relations. Interviewing such groups in El Paraíso and Santa Bárbara indicated that while many were struggling to make positive net incomes from maize, and some had difficulty in supplying consumption needs, one or two had partially consolidated their activities on a relatively secure financial footing and were able to sustain maize farming, make a cash income and provide adequate maize for consumption, with the possibility of continued access to credit to finance seasonal cash outlays (see Table 6.2). For the remaining groups, reproducing maize was more or less precarious. Furthermore, some struggling groups exhibited semi-proletarian characteristics by doing wage work to survive. There were key differences between collectively organized groups' capacities to produce and reproduce maize production and consumption and those of semi-proletarian farmers: (i) they had more or less assured access to land; (ii) access to land was independent of exchanges agreed with local landowners; (iii) groups had political or ideological principles which guided internal organization and distribution of resources and output as well as social relations with outside individuals and agents; (iv) collective organization also provided internal solidarity and cohesion; (v) groups could take joint action in relation to external forces and pressures. This is not to idealize collective organization or to ignore the problems faced by many such groups. These findings do however suggest that collective action can help reduce farmers' vulnerability.

Credit

An important policy area in the 1980s was the delivery of institutional credit to small farmers. Government intervention was largely carried out through Integrated Rural Development Programmes, or DRIs. The DRIs carried out a range of activities to promote rural livelihoods, but production credit was the foundation for much of their operations.

Small farmers such as those semi-proletarian maize farmers discussed above, did not generally receive institutional credit. Farmers who fell

Table 6.2 Number of collectively-organized groups with guaranteed access to land and ability to reproduce maize among those interviewed in El Paraíso and Santa Bárbara, 1987 and 1988

	All groups	Guaranteed land	Members carried out wage work	Received institutional credit	Made pre-harvest sales	Made positive net income	Sufficient maize for consumption after sales
El Paraiso	4	4	1	3	1	3*	2/3***
Santa Barbara	10	10	1	8	Only individual members	6**	5**

* One group had very low income.
** Data known for 6 groups only.
*** One group had an outstanding debt and would have had to sell maize needed for consumption.

within the remit of DRI credit programmes were generally petty commodity producers (as well as some collectively-organized groups). However, my data show the potential to transform the production of such farmers could be undermined by indebtedness.

The models used by the DRI in El Paraíso to assess credit needs of petty commodity producers assumed given quantities of purchased inputs and use of waged labour. In practice, farmers tended to use fewer inputs than in the model (even with credit) and supplied part of the labour themselves or from family members. Thus they were often being allocated relatively substantial sums of money which could be difficult to repay at harvest-time from their maize sales. There were two results: farmers fell into debt; maize needed for consumption was sold. Of 12 farmers interviewed who were receiving credit from the DRI, 5 could repay their loans from maize sales (and have enough maize for direct consumption); 3 would have been able to repay their loans if they had saved the difference between the loans and their actual cash costs (in practice, it was more likely that the money was used as a consumption fund during production); and 4 could not repay their loans by any of these means (Table 6.3).

These data illustrate some of the problems with such credit programmes. They show how policies intended to improve rural livelihoods can help to create vulnerability and undermine farmers' entitlements. First, recipients of rural credit were actively encouraged to commoditize their production processes as well as commoditizing their output. A number of farmers were left in difficult and contradictory situations because, on one hand, they needed credit to continue to purchase inputs and hire machines, while, on the other hand, repaying the amount of loan received was not

Table 6.3   Ability of farmers receiving DRI credit to repay loans from maize sales, El Paraíso, primera 1986–7

Farm size group (Has)	Number interviewed in group	Could repay loans plus interest from total maize sales from total	Could repay loan plus interestif saved difference between loan and cash costs	Could not repay loan by these means
< 1–5	4	1	1	2
5–50	8	4	2	2
Total	12	5	3	4

always possible from maize sales and debt relations could jeopardize future access to credit. Second, although institutionalized loans were apparently based on freely contracted agreements involving specified repayment by a particular time, the relationship of petty commodity producers to credit schemes also involved an implicit power relation with the creditors in which agreements about the use of resources might be against farmers' better judgements. Moreover, inclusion and exclusion from credit schemes might involve other issues than financial need and credit-worthiness, such as bureaucratic decision-making or political influence.[9]

Output markets

A third area of exchange which can create or reinforce vulnerability to food insecurity for maize producers is output markets. There was a growing literature in Honduras in the 1980s about the functioning of grain markets, state intervention and the role of prices in stimulating or discouraging production for the market.[10] On one hand, there was considerable concern about the role of the Honduran Marketing Board (*Instituto Hondureño de Mercadeo Agrícola* – IHMA) and the extent to which its interventions created market 'inefficiencies'. (In practice, the majority of maize farmers sold to private traders because of the high transaction costs in selling to the IHMA (Loria and Cuevas, 1984, p. 24), and the preference given to those receiving institutional credit (p. 40). On the other hand, there was continuing reluctance among many policy-makers in the 1980s to reduce the activities of the IHMA and to deregulate trade (including imports and exports) and prices.

Policies to encourage increased output through price incentives and promote market efficiency need to take into account the different types of exchange engaged in by farmers, and the different conditions under which they take place. Debates in the 1980s sometimes assumed that small farmers (under 5 hectares) only produced maize for consumption (Aguirre and Tablada, 1988). It was also often assumed that farmers generally engaged with maize markets in conditions of competition.

Although some of the farmers I interviewed engaged in forced commerce (Bhaduri, 1983) through pre-harvest sale arrangements, and others were obliged to sell maize needed for consumption to repay institutional debts, there was no doubt that all farmers engaged with markets for maize. Among 28 individual farmers (semi-proletarian, petty commodity and commercial) interviewed in El Paraíso, there was no-one who sold less than 30 per cent of their output. Most, including semi-proletarian farmers, sold more than 50 per cent of their harvests and several sold nearly all they produced, even if they were left with consumption deficits.

Even in apparently competitive markets, there are different mechanisms of ensuring control over supplies, sales and profits (Johnson, 1995). Particularly important for maize farmers' market entitlements is how the social relations of trade function locally. For example, although one study of such relations (Loria and Cuevas, 1984) estimated that (only) 24 per cent of sales by farmers was made to nearest or only buyers. Loria and Cuevas concluded that trade was competitive but these data could also suggest that there was considerable local control over markets.

I cannot draw clear conclusions from my own data, but the following observations suggest that local social relations can affect the functioning of markets and prices. First, the origin of maize purchasers indicates that the marketing networks of semi-proletarian farmers were relatively confined. Comparing type of farmer interviewed in the department of El Paraíso with the origins of traders to whom they sold their maize (including pre-harvest sales), I found that just under half of semi-proletarian farmers made sales to local purchasers (same village), and that the majority of petty commodity producers and commercial farmers sold to traders coming from the market centres of Danlí (El Paraíso), Tegucigalpa (capital city) and Choluteca (Southern Honduras).

Second, many local traders were also commercial farmers (including those who rented out land). Some made pre-harvest purchases. Although these activities could result in many possibilities for agreements between farmers and traders, and therefore competition, commercial maize farmers who engaged in local trade as juntadores (agents who stockpiled locally-produced maize), and then sold it on, needed to ensure that prices they paid to local farmers were lower than those they received from other purchasers.

Third, relating the origin of traders to prices paid to farmers suggests that higher prices were paid by traders from the three main market centres than by local purchasers. In Table 6.4 Lps. 34/Kg was the average price received by semi-proletarian farmers when pre-harvest sale prices were removed. It can be seen that the higher prices comprise entirely sales made to traders who came from *outside* the local villages.

Did this mean that semi-proletarian farmers generally received lower prices than other farmers? The evidence from my fieldwork is that the most important element in price was time of sale, which particularly affected semi-proletarian farmers. Although all farmers were under pressure to sell at least part of their output after harvest to repay debts, the earliest sales of all (and those which were pre-harvest) were made by semi-proletarian farmers. However, it was also found that the collectively-

Table 6.4   Prices received by individual maize farmers interviewed in El Paraíso and origin of purchasers (primera 1986–7)

Lps/Kg	Same village	Nearby village	Danlí	Other in El Paraíso	Tegucigalpa	Choluteca
.17	3					
.22				1		
.28			1			
.30	1				1	
.32						1
.33	1	2	2			
.34				1		
.35					1	1
.36			1		1	
.37					1	
.38			1		1	
.39			1		1	
.40			2			
.41					1	
.42				1		
.44			1		1	2
.46					2	

organized farmers in El Paraíso generally obtained somewhat higher prices for their output than the average prices received by semi-proletarian farmers in the same area. Their average time of sale was also later. These data suggest that collective organization may have enabled them to withstand pressures to make very early sales (Johnson, 1995, p. 329–30).

4   POLICY DILEMMAS FOR THE 1990S

The late 1980s and early 1990s have seen the growth of economic reactivation and structural adjustment policies in Honduras, including liberalization of markets. Among markets affected are those for land, credit and output, for which some implications of the above discussion can be drawn, even if of a somewhat speculative kind.

Land

A key issue for any land policy in Honduras is the highly unequal distribution of land and the lack of access to land for many rural people. Although land redistribution effected by the 1975 land reform law was partial, a considerable number of landless and semi-proletarian farmers secured means of production. Moreover, the experiences of the collectively-organized maize producers referred to in this chapter suggest that collective action on land was able to improve on the positions of semi-proletarian farmers, even if difficulties were also apparent.

Security of tenure is also a key issue. However, the current titling programme is unlikely to provide security for small producers such as semi-proletarian maize farmers, who often rent or borrow very small plots of private land (or land controlled by other farmers), although it could reinforce the positions of petty commodity producers and help to consolidate commercial farms with mixed forms of land tenure. Furthermore, although one of the measures in the legislation is to facilitate the renting of land, the reinforcement of land markets could result in increased rents and land values, putting prices outside the reach of many petty commodity producers as well as semi-proletarian farmers. Such developments are likely to be reinforced if the importance of returns to land and land productivity grows with attempts to increase the commoditization of land. Export crops and other high value food crops are likely to expand, in so far as there are markets for output. If the value of land increases sufficiently, landowners may also gain more from buying and selling land than from investing in production.

Credit

There are many problems in applying rural credit programmes in contexts of inequality. On one hand, access to credit can free farmers from some of the social relations of dependence on patrons, whether wealthier family members, neighbours or commercial farmers. On the other hand, the way credit is calculated and allocated can result in high loans compared with farmers' capacities to repay them, especially if farmers experience high pre-harvest crop losses.

Within the structural adjustment programme of the late 1980s and early 1990s in Honduras, interest rates have been liberalized and the costs of credit have increased (Thorpe, 1992, pp. 150–1). This would affect petty commodity maize producers and commercial farmers directly, but could also affect the terms on which semi-proletarian farmers and some petty commodity producers made personalized loans, especially if their creditors depended on institutional sources of funds. This increased vulnerability combined with the short-comings of earlier credit programmes suggest that participative schemes in which farmers can help to define the parameters and rules of loans are urgently required.

Output markets and pricing

Social processes need greater consideration in policies designed to make the functioning of markets more efficient and in decisions whether to regulate or deregulate prices. Price effects can only be fully analysed and evaluated by acknowledging the non-commoditized and personalized aspects of exchange, and the role of social hierarchies and power. In addition, the personalized exchange relations identified in this analysis rely on commoditization and can therefore be affected by price changes.

These considerations are not immediately evident in the reform of output markets which has taken place since my study was carried out. Domestic prices for agricultural products have been liberalized within Honduras's structural adjustment programme, although some, such as grains, have been subject to intervention if they fall outside price bands. The rationale behind price liberalization was to increase agricultural output, based on the assumption that prices would rise and provide incentives to farmers. Market prices for maize did rise (Thorpe, 1992, p. 144), in principle benefiting petty commodity producers and commercial farmers (assuming costs of production did not experience equivalent increases), but undermining net incomes for semi-proletarian farmers and any petty commodity producers who had to purchase maize for household consumption.

Food security as a sustainability issue

This analysis has demonstrated some of the points of vulnerability of semi-proletarian and petty commodity maize producers in Honduras. Although I have looked at a single crop, this type of analysis can be applied to other areas of small farmer production. These points of vulnerability pose real dilemmas for policy. On one hand, they constrain farmers' capacities to diversify production (or engage on other forms of income generation) as well as increase maize output. On the other, the alternatives outside small-scale farming are limited, hence many farmers' pursuit of strategies to reproduce their existing levels of production, even if they depend on subordinate relations with commercial farmers and traders.

Sustainability as a process of change to build human capacities and to create a resource base to meet human needs may thus suggest that such production forms are not tenable, and that the critical issue for Honduran small farmers is the creation of other forms of employment. However, the commercialization of agriculture based on high value crops is likely to lead to further displacement of small producers without necessarily creating an adequate employment base. To be able to predict the outcomes of such (or other) policies requires understanding the structural basis of inequality in the countryside, the complex social relations which reinforce it and the types of action which might challenge it and improve the quality of life rather than 'sustain' poverty.

Notes

1. This has been a source of debate; see, for example, that between Bernstein and Long (1986).
2. Fictive kin is the attribution of kinship relations to people who may not be blood relatives but who have rights and responsibilities similar to family ties.
3. The notion of 'semi-proletarian' is problematic if one takes the 'semi' literally. I use the term here to denote those small farmers who are engaged in systematic wage work as well as production on their own account.
4. The early steps towards titling concentrated on coffee farmers and those with a minimum of 5 Has of land, much less land than any semi-proletarian maize farmer would have; the minimum was later reduced to 2.5 Has, also more than that generally accessed by such farmers.
5. 'He lends/gives/rents me a manzana [.7 Ha].'
6. These farmers may also have worked for other landowners than the one who rented them land.
7. First and main maize harvest.
8. Pre-harvest sales were sales of maize at fixed prices made early in the crop cycle to finance fertilizer and other inputs. Loans made in this way were usually repaid in maize at harvest time and incurred an interest rate of about 100 per cent as prices paid by the creditor were generally half the market price.
9. Extensionists stated that decisions about who should receive credit were made by the central administration of the programm – recipients were often those who looked best on paper. There was also concern that credit was being allocated to farmers who had more land than specified in the criteria and/or who had other sources of income.
10. For example, Aguirre and Tablada, 1988, 1989; Economic Perspectives, 1986; Larson, 1982; Loria and Cuevas, 1984; Norton and Benito, 1987; Pollard *et al.*, 1984; Quezada and Scobie, n.d.

Bibliography

Aguirre, J.A. and Tablada, G. (1988) *Macro Análisis de la Producción de Granos Básicos en Honduras 1976–87*, Instituto Interamericano de Cooperación para la Agricultura (IICA), Tegucigalpa.
Aguirre, J.A. and Tablada, G. (1989) 'Nivel Tecnológico y Tasas de Cambio y su Efecto sobre la Protección Nominal y Efectiva: Implicaciones para el Dise/no de la Políticas de Investigación en Granos Básicos, con Enfasis en Maíz en Honduras', Paper presented to the XXXV Annual Meeting of the PCCMCA, San Pedro Sula, Honduras, April 3–7.

106   *Lessons from Honduran Maize Farming*

Austin, J.E. and Esteva, G. (eds) (1987) *Food Policy in Mexico. The Search for Self-Sufficiency* (Ithaca and London: Cornell University Press).
Bardhan, P. (1991) 'On the concept of power in economics', *Economics and Politics*, Vol. 3(3), pp. 265–77.
Bartra, R. (1976) *Estructura Agraria y Clases Sociales en México* (Serie popular Era), Instituto de Investigaciones/UNAM, Mexico.
Bhaduri, A. (1983) *The Economic Structure of Backward Agriculture* (London and New York: Academic Press).
Bharadwaj, K. (1985) 'A View on Commercialization in Indian Agriculture and the Development of Capitalism', *Journal of Peasant Studies*, Vol. 12(4), pp. 7–25.
Brass, T. (1986) 'The Elementary Strictures of Kinship: Unfree Relations and the Production of Commodities' in Scott (ed.).
Brockett, C.D. (1990) *Land, Power and Poverty. Agrarian Transformation and Political Conflict in Central America* (Boston: Unwin Hyman).
Drèze, J. and Sen. A. (1989) *Hunger and Public Action* (Oxford: Clarendon Press).
Economic Perspectives Inc (1986) 'Una Evaluación del Instituto Hondure/no de Mercadeo Agrícola (IHMA)' USAID, Tegucigalpa.
Gabaldón, A. J. (1992) 'From the Brundtland Report to Our Own Agenda', *International Journal of Sociology and Social Policy*, 12(4/5/6/7), pp. 23–39.
Hewitt de Alcántara, C. (ed.) (1993) *Real Markets: Social and Political Issues of Food Policy Reform*, London and Portland, USA: Frank Cass, EADI and UNRISD.
Janvry, A. de (1981) *The Agrarian Question and Reformism in Latin America* (Baltimore and London: The Johns Hopkins University Press).
Johnson, H. (1995) *Reproduction, Exchange Relations and Food Insecurity: maize production and maize markets in Honduras*, PhD Thesis (Milton Keynes: Open University).
Larson, D.W. (1982) 'The Problems and Effects of Price Controls on Honduran Agriculture', Ohio State University: department of Agricultural Economics and Rural Sociology, Economics and Social Occasional Papers, OSU No. 929.
Long, N., van der Ploeg, J.D., Curtin, C. and Box, L. (ed.) (1986) *The Commoditization Debate: labour process, strategy and social network* (Netherlands: Agricultural University Wageningen).
Loria, M. and Cuevas, C.E. (1984) 'Basic Grains: Marketing Channels and Financing at the Farm and Wholesale Levels', Ohio State University: Department of Agricultural Economics and Rural Sociology, Agricultural Finance Program.
Mackintosh, M. (1990) 'Abstract markets and Real Needs', in H. Bernstein and B. Crow.
Mackintosh, M., and Martin, C., *The Food Question: Profits versus People?* (London: Earthscan).
Munslow, B., Fitzgerald, P. and Mc Lennan, A. (1995) 'Sustainable development: turning vision into reality' in Fitzgerald *et al.*, *Managing Sustainable Development in South Africa* (Oxford: Cape Town).
Norton, R.D. and Benito, C.A. (1987) 'Evaluación de los Programas Realizados bajo el Título I de la Ley Pública 480 en Honduras', Tegucigalpa: USAID, Oficina de Desarrollo Rural, September, mimeo.

Oswald, U. (ed.) (1979) *Mercado y Dependencia* (Mexico: Editorial Nueva Imagen).

Pollard, S.K., Graham, D.H. and Cuevas, C.E. (1984) 'Coffee and Basic Grains: a Review of Sectoral Performances, Pricing and Marketing Margins and Recent Policy Changes', Ohio State University: Department of Agricultural Economics and Rural Sociology, Agricultural Finance Program.

Quezada, N. and Scobie, G. (n.d.) 'La Situación de Oferta y Demanda y Necesidad de Importaciones de Maíz en Honduras', English Executive Summary (prepared by R. Franklin), Sigma One Corporation.

Scott, A. MacEwen (ed.) (1986) 'Rethinking Petty Commodity Production', Special Issue, *Social Analysis*, No 20, Adelaide.

Sen, A. (1981) *Poverty and Famines: An Essay on Entitlement and Deprivation* (Oxford: Clarendon Press).

SIECA (1984) *Modelo Sectorial de Programación Lineal para la Producción Nacional de Granos Básicos*, Tegucigalpa.

Thorpe, A. (1992) 'Caminos Políticos y Económicos hacia la Reactivación y Modernización del Sector Agrícola. Una vez entendido el Ajuste' in Noé Pino, H., Thorpe, A. and Sandoval Corea, R., *El Sector Agrícola y la Modernización en Honduras* CEDOH/POSCAE, Tegucigalpa.

WCED (1987) *Our Common Future*, The Brundtland Report (Oxford: Oxford University Press).

Wuyts, M., Mackintosh, M. and Hewitt, T. (1992) *Development Policy and Public Action* (Oxford: OUP in association with the Open University).

7 Diversity and the Nature of Technological Change in Hillside Farming in Honduras
Kees Jansen[1]

1 INTRODUCTION

Honduras has a good deal of hillside agriculture which is thought to be unsustainable because it provokes soil degradation and erosion. Nevertheless producers continue to use slopes in many different ways and this raises the question of how Honduran producers have responded to the challenge of mountain ecology in the context of ecological, social and economic change in a particular region.

In this chapters I refute the classical dualist image of Honduran agriculture as being composed of a technologically advanced, export oriented, capitalist agriculture in the valleys *vis-à-vis* a traditional production system of the peasantry on slopes. An empirical description is presented which shows that the variety of production systems in a marginal hillside area is much wider than could be explained by a dualist view. The municipality El Zapote in North-West Honduras (6000 inhabitants in 1990), is used as a case. It is located in a mountainous region of Honduras where the main sources of livelihood are the cultivation of coffee, tule (a fibre crop), maize, and beans, and keeping of cattle. Data have been obtained from formal fixed interviews with 83 producers and many farm visits between 1992 and 1994, and are compared with evidence from statistical sources, the municipality archive, and a survey I conducted.

After reassessing some theoretical notions on diversity and change, the third section deals with producer responses to biophysical production constraints. Then I examine how technology in different crops and grasslands has changed during this century. In a subsequent section I explain mechanism behind the described technological changes.

2 CONCEPTUALIZING DIVERSITY AND TECHNOLOGICAL CHANGE

Change in agricultural production has been studied from so many approaches and disciplines that a thorough review in a few pages would be far too ambitious. To illustrate my own approach I refer to some recent studies on the technical organization of agricultural production.

Until recently, traditional agriculture and modernized agriculture were considered to be essentially different. This view permeates many approaches from the political right (studying innovation) to the political left (studying exploitation in agrarian structures) in policy and social sciences. Basically, the 'traditional' is subsistence agriculture with simple tools, a mixture of crop cultivars, low external input level, etc. The 'modern' is more capitalized agriculture with wage labour relations, monocropping, high energy use, high external input, direct links to input and output markets, use of high-yielding cultivars, and so on. This dichotomy is seen in relations of production as well as in technology, cf. Faber's use (1993) of the concept of functional dualism. In my view this dichotomic view continues to inform policy and extension in Honduras, notwithstanding research efforts to characterize differences in production systems (for example, Galvez *et al.*, 1990). Policies built on this view tend to transform the 'traditionally', and therefore negate the dynamics of technological adaptation, of change, by those producers who are considered to produce 'traditional' or with a 'low technology level'.

Within agronomy and rural sociology there have always been stances which nuance or deconstruct this dichotomy. Two examples are 'farming systems research' and 'styles of farming' research. The former tries to understand variation in production systems by discerning agro-ecological zones (areas of similar soil, vegetation, climate and population density). Because of the variation in production conditions one cannot assume that there is one optimal farm size and technology level within such zones. Within these zones producers were assumed to adapt to production constraints in the same way (for a review see Brouwer and Jansen, 1989). 'Styles of farming' studies take up the issue of diversity in production systems. They only criticize the idea of a dichotomy between 'traditional' and 'modern' but also the idea of a linear relation between levels of technology and output (Van der Ploeg, 1990). Furthermore, it is thought that producers do not react uniformly to similar environments but, instead, develop different actor strategies.

Diversity is thus approached by two types of abstractions; the former points to the biophysical production conditions as the causal factor, while the latter focuses on variety in actor strategies. What is under-theorized in both these approaches is how biophysical variation, actor strategies and structural socio-economic changes together convert production systems in an interacting process over time. A theory of diversity, therefore, should accept the idea that each producer varies in his involvement in a set of relations and production circumstances. This is opposed to building the theory on the possibility of producers being in a situation under exactly the same structural circumstances which only afterwards diverge because of agency. It also puts more emphasis on historical processes; for example, of social differentiation.

3   ECOLOGY, PRODUCER STRATEGIES AND DIVERSITY: SOME EXAMPLES

The rainy season in El Zapote in the north of Honduras lasts from May to December with two major peaks in June and September. Variation in rainfall (annual mean is 1615 mm) influences producers' practices more strongly than temperature (annual mean is 24.4°C). Both drought and excessive rain can be expected to occur regularly. In some years excessive rainfall encourages fungus infestation in the *milpa* (*milpa* refers here to the first maize crop, *postrera* stands for the second crop). The period until mid-June is critical because of possible drought stress in crops. The drier period from mid-July to mid-August (the *canícula*) is less marked than, for example, in the south of Honduras (Zúniga, 1990), but drought stress can occur. The rainy season is long enough for two maize cropping seasons, but there is a large probability that the *postrera*, (sown in November) will have shortages in water supply at the beginning of the dry season. To a large extent, the *postrera* depends on the water-holding capacity of the soil and the particular field, and is threatened when soils lose their water-holding capacity through, for example, erosion or decreasing organic matter content. The dry season causes problems for grassland management, especially on steep slopes with shallow soils with low water retaining capacity.

Although most soils in El Zapote have developed on limestone, different zones can be discerned (partly an effect of local climatic variation because of altitude differences). The zones north and east of El Zapote, used for maize, beans, and grassland, show deficiencies in potassium and sometimes in phosphorus. Nevertheless most of these fields are free from stones. In contrast to the mountain zone, soils in the east are less acid (mean pH of 5.7) and have high levels of calcium and magnesium. 'The mountains', to the south-east of El Zapote, have a markedly higher annual rainfall than the lower parts. Mean temperatures here are lower than in El Zapote, because this zone ranges from about 900 to 1600 m.a.s.l. Precipitation can be expected during the dry seasons and the rainy season starts earlier. The soils in the mountains are extremely or very strongly acid (pH range 3.4–6.3; mean 4.3, values below 4.0 are associated with high levels of aluminium). The zone to the south has soils which 'dry out' very rapidly and there is a high risk of losing the harvest on them in dry years. When sowing, patches of soil have to be sought out between the numerous stones and rocks. The village commons, to the west of El Zapote, are predominantly used for maize cultivation; bean plants do not form pods or seeds. Here too there are stones and rocks. On a limited area two rivers have formed river terraces with alluvial soils. It can be concluded that soil conditions vary over short distances. For instance, the village commons are not suited for beans, but only 500 m to the north are the favourite fields for bean cultivation. Furthermore, texture within one field can vary from heavy clay to sand.

The following examples illustrate how variation in production systems is related to producers' responses to these physical and sociological conditions. Firstly, many producers aspire to acquire different fields in different ecological zones for different objectives (different crops), such as in the mountains for coffee, on the river terraces for tule, on plots with water sources near the village for grasslands for pack animals, in northern zones plots which do not dry out fast for maize, and so on.

Secondly, producers phase their planting in time and space. This is a major strategy to enlarge the variation in crops and to augment their yields. For example, producers sow vegetables and beans during the dry season months December–February on the river terraces, on soils with a large water-holding capacity and enough water for the growth period of the crop. In this season less fungal infestation occurs. These fields, and the fields sown in the mountains between February and April, provide good quality seed for the two main bean seasons (June–August and September–October). Although less pronounced, the same strategy can be observed in maize cultivation. Maize can be sown on some river terraces until late December. Geographical and temporal distribution of crops is partly in accordance with a spread of labour input. However, to refer solely to the labour issue would be reductionist, as it is concomitantly an adaptation to biophysical factors. The type of soil and its water-holding capacity determine the period which is optimal for sowing.

Thirdly, yield expectations for *milpa* and *postrera* result in fewer producers sowing *postrera* than *milpa* (80 per cent versus 96 per cent), and in the area sown to *milpa* per producer being larger (0.86 ha) than that sown to *postrera* (0.56 ha). Producers expect *postrera* yields to be half those of the *milpa*. This coincides with my survey data. The mean *milpa* yield was 825 kg/ha (1994) and for the *postrera* only 428 kg/ha (1993). Those producers who continue to sow *postrera* give it less attention than the *milpa*. Fewer producers apply fertilizers to their *postrera*, which indicates that producers take less investment risks for the *postrera*.

These examples of strategies of coping with ecological constraints do not lead to uniform strategies and an optimal use of ecological zones. Firstly, the household needs and opportunities mean that every producer takes different decisions. For example, a farmer with only a small plot of land will try to produce for his most urgent needs (normally maize), notwithstanding local perception of land suitability. His needs will override the optimal use in an ecological sense. Secondly, the availability of land for a producer is socially, economically, culturally, and politically regulated. Thirdly, the ecological constraints are not 'outer limits', immutable natural circumstances. Winds have always blown, but wind becomes increasingly damaging when the surrounding forest is cut and can no longer function as a wind break. Sowing *Planta Baja*, a new maize cultivar with short stalks, is the farmer's response to this problem. With their agricultural practices people continuously interfere with nature. With changes in technology, their conception of the main ecological constraints will alter. In the following section the main technological changes in El Zapote will be outlined.

4   PATHS OF TECHNOLOGICAL CHANGE

In this section I will discuss how technological practices in the main crops in El Zapote have changed during this century. My analysis corroborates other recent studies (Netting, 1993; Brush and Turner, 1987) which show that peasants are masters in organizing adaptation to new circumstances and technologies.

Maize

Most producers in El Zapote cultivate maize for household consumption. Maize is sown with a dibble stick, without any soil tillage. Fungal diseases are the most important threat to maize apart from drought. Pest damage is less critical. The following description of different production systems is more extensively discussed in Jansen (1995a).

During the first decades of this century producers combined a Low altitude forest fallow system with a Mountain forest fallow system. Primary or secondary forest was felled with axes. Small trees were used to build fences. The harvest was stored in small huts in the fields. Generally, only a *milpa* was prepared. Only in case of a poor first harvest was the *postrera* sown. The low altitude system was sometimes combined with *milpa* in a mountain forest fallow system, on fields at altitudes above 900 m. Mountain agriculture could guarantee maize production in dry years when *milpas* in lower areas did not yield, despite fungal infestation in normal years because of high humidity at this altitude. Both systems existed under low pressure on land. Cattle, pigs and pack animals grazed freely in the areas under fallow.

Several producers started to accumulate cattle and began to fence tracts of land at lower altitudes at the end of the 1940s and in the 1950s. In that period a skewed distribution of land was established. These producers employed a Forest fallow-maize-pasture system. Tenants cleared the land and could cultivate maize during 1 to 3 seasons, sometimes paying a rent in kind of 1.5–2 loads of corncobs per *manzana*. During the cultivation period grass infestation started. The land was grazed by cattle afterwards, while tenants moved on to another plot in the forest. The land with grasses reverted to secondary forest after a few years. During the 1960s this system gradually changed towards a system in which forest was converted into permanent pasture without any fallow period.

The establishment of permanent pastures by expanding cattle owners, as well as an increase in the number of producers, reduced access to land. In this context a Bush Fallow system developed, characterized by a shorter fallow period (6–10 years). The axe was no longer the most important tool; instead, the importance of the *machete* and the *pando* (for weeding) increased.

The use of fallow and burning to clear the fields were two main practices to maintain soil fertility. With the emergence of bush fallow systems a reduction of the fallow period was combined with a continuation of burning. This generated severe problems with soil fertility and thus with yields. Two responses to this problem were an expansion of the *postrera* and an extension of cultivated lands.

Pressure on land, however, continued to increase, as did the demand for maize, and the fallow period continued to be shortened. These production constraints led to the emergency of an amalgam of different practices rather than a single new production system. Technical changes that occurred were related to burning (or not burning), the disappearance or reintroduction of a fallow period, use of fertilizers and herbicides, cultivars, plant spacing, storage and pesticides. Some of these changes were 'imported' while others were generated locally (adaptations).

The first use of chemicals started in the 1950s with small quantities of chlordane being applied in maize storage in the house. Herbicides, especially paraquat, became popular in the 1980s and were distributed via local traders, in a period that weed infestation became a pressing problem for producers as an effect of shortened fallow periods and frequent burning. Fertilizer use started after 1985. Fertilizer application masks decreasing soil fertility.

New cultivars were introduced by producers who returned after temporarily migrating to the north coast. The famine year (probably 1954 or 1955) forced people to eat their own maize seed, and they therefore had to obtain seed from elsewhere, often from new cultivars (*Maizón, Tusa Rosada* and *Capulín de Estíca*). During the 1980s development programmes introduced high-yielding cultivars such as *Planta Baja, Guayape* and *Hibrida*.

Burning has remained an important practice; however, many producers have abandoned this practice.

The classical way of sowing maize is *en cuadro*, a square pattern of holes, with 3–5 seeds put in each hole. A new spacing pattern has emerged, in which maize is sown along the contours. The new pattern is *en surco*, with the holes in a line, 2–3 seeds per hole and more holes per unit area.

It was noted above that the *postrera* became more important in general. However, nowadays several producers have said that they intend to stop sowing *postrera* because of increasing drought stress. Others are experimenting with an earlier second sowing, in August instead of in November, *postrera de agosto*.

Occasionally, producers try to re-introduce a fallow period, for which they need to acquire more land. Those with sufficient sources of income (coffee) can aspire to this.

Another important change to take place concerns storage of maize. The huts in the fields have disappeared, as has cob storage in the house. In general, the kernels are now stripped from the cob and dried in the sun before storage. The grain is put into bags, old oil drums or in special silos containing just under 1000 kg grain, and stored in the house.

These technical changes are not abrupt leaps from the old technique to a new one. Instead, for every variable different values can be noticed. This corresponds with a diversity of ways of carrying out production. Hence, the dual idea of the co-existence a modern, high input, production system next to a traditional low input system is inappropriate. It is not possible to distinguish two such systems (traditional versus modern technology) in this village. Instead, all technological combinations can be considered as contemporaneous because producers apply them with their own calculations and perceptions of actual existing opportunities and constraints, social relations, and so on. None of these combinations looks like those common a few decades ago. A characterization which classifies technological systems into low, intermediate and high input might do more justice to differences in technology. However, such a classification also encounters several problems.

The definition of levels of technology is problematic when reviewing more than a few simple variables, especially if we wish to acknowledge the diversity originating in ecological variation. Furthermore, technology is not one package. Producers can apply high levels of fertilizers, combined with burning, while not using herbicides and sowing high-yielding cultivars. Is this a low technified system, a moderately technified system, or a technified system? A related problem is that technology is mostly only considered as artefacts: ploughs, tractors and fertilizers. This is more easy to conceptualize than, for example, burning or not burning, spacing, intercropping, green manure, compost, times of weeding. Under different circumstances the definition of what is more technified and what is less technified can become very elusive. Another problem is that the policies which are based on classifications according to level of technology often reintroduce the idea that low level technology is backward and traditional. Subsequently, the main target is how production can be transformed from a low input level to a high input level, without feeling the need to analyse further the nature of differences.

Coffee

Recently, the nature of the expansion of coffee in Honduras has become subject of debate (Baumeister, 1990; Jansen, 1993). In this section I will only review the main technological changes in coffee cultivation that have accompanied this expansion in El Zapote. Average yields increased from 195 kg/ha in 1952 to 862 kg/ha in 1992 (DGECH, 1954; SECPLAN, 1994).

At the beginning of this century, rows of undergrowth in the forest were slashed with axes and machetes, and coffee seedlings were planted directly with a dibble stick. There were differences in how strictly producers planted in rows. This way of cultivation changed in the 1960s, when the first seed beds and nurseries were made. All the vegetation in the forest plot was felled and burned before the seedlings were transplanted. The holes for transplanting were still made with a dibble stick. This changed in the second half of the 1980s when bigger holes became the rule, accompanied by a much denser spacing. Since then, burning has become less common when clearing fields for planting coffee. These changes also led to changes in the shade vegetation. Initially, the producer maintained all tall trees and the vegetation displayed several storeys. Later, increased attention was paid to selecting trees to be retained for shade, and fruit trees were occasionally planted. Clear felling became common practice in the 1980s, after recommendations of extensionists. The sudden removal of shade led to disastrous effects and several coffee groves were lost. Later, producers again selected desired shade trees (*Inga* spp.) by not weeding them from the sprouting vegetation. Only recently, some producers have started planting most of the desired shade trees (including bananas).

In time, more different coffee cultivars were introduced. The cultivar *Arábigo* (other local names are *Indio*, *Café del País*, and *Fantasía*) was supplemented with the cultivar *Bourbon*, which had spread northwards from large plantations in El Salvador in the 1950s. Both are tall cultivars. In the 1980s shorter, high-yielding cultivars were planted: *Caturra*, *Catuai* and *Paca*.

Various systems of weeding have been employed. Initially, fields were only cleared of weeds and bushes with a *machete* just before the harvest. This developed into a system in which weeding is carried out two or three times with the *pando*. The form of harvesting also changed, to three picking rounds instead of just one. Producers use herbicides incidentally. These producers do not always have the most input-intensive systems. They do not opt to hire day labourers for weeding, or they have much grass in their fields which cannot be eradicated with the *pando* alone.

Some of these technical changes interacted with shifts in fungal diseases. In systems with unrestricted shade vegetation the most serious disease was 'ojo de gallo' (*Mycena citricolor*). When nursery systems and reduced shade became common practice, other diseases emerged: 'damping-off' in nurseries (*Rhizoctonia solani*) and 'mancha de hierro' (*Cercospora coffeicola*) in spots with much direct sunlight. In the early 1980s coffee leaf rust (*Hemileia vastatrix*) reached El Zapote and fiercely attacked the fields at lower altitudes. As a consequence, coffee was planted at higher altitudes, where the diseases 'derrite' (*Phoma costarricensis*) and 'antracnose' infested coffee. The emergence of coffee leaf rust was, furthermore, contested with fungicides, mainly copper sprays. These also reduced attacks by other fungi.

Many producers purchased knapsack sprayers to combat coffee leaf rust in the 1980s. This also stimulated herbicide spraying and the use of endosulfan against attacks of coffee berry borers. When coffee producers replaced forest vegetation at higher altitudes with groves with few shade trees, the temperature near ground level rose and coffee leaf rust and the coffee berry borer spread to higher altitudes. Not only pesticide use but also fertilizer application has increased.

Coffee processing has also undergone to technical change. In the first half of this century coffee was sold as dried berry or clean coffee. From the 1960s onwards, wet processing became common and nowadays all producers themselves use hand pulpers (often hired) to remove the beans from the berry.

Beans and tule

Technological changes in beans and tule have not been as common as in maize and coffee. Beans are sown as a sole crop on small plots by 67 per cent of the respondents in the survey (mean area 0.15 ha; 800–900 kg/ha is considered a good yield). Most beans are sown for household consumption. Climbing beans are intercropped with maize. Bean cultivation has undergone a change in spacing. A greater density of holes is somewhat compensated for by putting fewer seeds (1 or 2, sometimes 3) in each hole; producers report that in the past they used up to 5 beans per hole. Producers explain this reduced number of seeds in terms of the diminished soil fertility.

Tule is a *Cyperaceae*. Its drying leaf stems are cut and woven into sleeping mats which are sold to traders. The crop grows on fields that never dry out. Stands can easily remain for 20 years. Although the total area planted to tule is small, the economic as well as the social and cultural importance of tule is paramount. Most women work with tule daily. A complex set of social relations exists between men, in general the owners of the tule fields, and women in their households, as well as between women who own tule and women who work with their tule on several different contractual arrangements (Roquas, 1994).

Cattle, pack animals and grasslands

Grasslands are important for cattle as well as for keeping pack animals. Most cattle (for milk and meat) is of the *Criollo* type, although some larger cattle farmers have crossed *Criollo* cattle with *Brahman*. Milk production does not exceed a few bottles (of 0.75 cl.) per day per cow. The dry season is an important constraint for dairy farming as it causes a shortage of quality forage. Horses and mules are used as pack animals; 56 per cent of the producers have one or more pack animals (mean of animals in this group is 2.0; range 1–7). Pack animals are used to bring the harvest of maize, beans, coffee and tule from the fields to the house.

Historical presence of cattle and pack animals

The increase in head of cattle during the last 60 years has been accompanied by a concentration of these cattle in the hands of fewer people. In 1952, 31 per cent of the producers had cattle, but by 1992 this had fallen to 11 per cent (the absolute number of producers with cattle remained constant). This concentration was probably even more important during the second half of the 1940s, when the so called 'free lands' came under control of individual persons.

The total number of pack animals has increased from 249 in 1952 to 569 in 1992. In the past, many of the pack animals were used by local traders to carry products to the city and to return with merchandise for their shops in the village. After roads had been constructed the pack animals were replaced by cars. Nevertheless, the number of pack animals continued to increase. Producers say that in the past there was less need for animals, as the maize and tule fields were closer to the village. Furthermore, there was no need to transport the maize harvest to the house in one go: instead, it was stored in the field. Occasionally a bag of maize was carried from the field to the house. Bringing the harvests to the house in one go without animals would be back breaking. Cross tabulation of my data shows that pack animals are associated with coffee production (statistically significant). The location of the coffee fields has shifted from near the village to far away in the mountains. The rise in coffee production has also contributed to an increased demand for pack animals. Thus, coffee expansion increased the demand for pack animals, and the revenue from coffee enables the animals to be maintained. Between 1952 and 1992 the number of cattle increased by 56 per cent from 726 in 1952 to 1131 in 1992. Total number of animals in this period, including pack animals, increased by 74 per cent.

Historical development of grasslands

During the first decades of this century pack animals, cattle and a few goats grazed on fields after the maize had been harvested, on land under fallow or under coffee, and on the 'free lands'. Free lands were all those lands which were not fenced or delimited. Producers could use these lands for annual crops, after officially submitting a request. Maize crops were temporarily fenced against the animals which foraged between these fields. The gradual disappearance of 'free lands', and thus of livestock cattle owned by the poor, has not yet completed the elimination of all animals owned by the poor. Roadsides and other fragments of land are still used for grazing pack animals, although local authorities regularly enforce the prohibition of grazing by impounding all roaming animals. On both the free land and the 'private' land, both cattle and pack animals grazed the maize stubble. Animals grazed fallow fields, if they could enter the dense secondary growth. Fallow has now become less important as a source of fodder. Therefore, for those without access to pastures stubble grazing remains the only option left.

The appearance of a forest fallow-maize-pasture system was discussed above. Between 1930 and 1950 the introduction of the grasses *jaraguá* (*Hyparrhenia rufa*) and *calinguero* (*Melinis minutiflora*) to this system changed the vegetation in a short period. Both grasses are very competitive with other species in a slash-and-burn culture. Producers started to weed woody vegetation (by using fire) to transform the system into permanent pasture.

Fencing was initially done to keep animals out of plots with crops. The fencing was temporary, and done with wood from the cut forest. Land was only used for two or three seasons. Over the years the function of the fence changed: from keeping animals outside the fence to excluding animals to keeping them in. When the latter function of the fence expanded, people started to talk of *potreros*, the fenced meadow. The disappearance of 'free lands' was accompanied by the appearance of more *potreros*.

The pressure on grassland (animals per unit area) increased by 22 per cent between 1952 and 1992, because the increase in number of animals (74 per cent) was accompanied by a much smaller increase in pastures, only 40 per cent. These figures fit in with data at national level presented by Howard (1989). In El Zapote, about 40 per cent of the agricultural land is now pasture, compared with 24 per cent in 1952. It must be pointed out that fields in fallow used to be a more important source of fodder when the fallow periods were longer. The use of pasture cannot only be equated with extensive cattle keeping; pack animals in 1992 make up about 33 per cent of all animals (26 per cent in 1952).

5 TOWARDS AN EXPLANATION OF CHANGE IN PRODUCTION SYSTEMS

An explanation of the technological changes in production systems cannot be limited to pointing to one (or two) determining variable or process. In my view, a more complete understanding combines several older theories about technological change, not by making an eclectic mix but by investigating the scope of different explanations. First, I will show that the growth of the *potrero* system (i) and the growth of coffee production (ii), took place in somewhat different processes than is suggested in literature about Honduras. Then I discuss the relation between population growth and the triad of fallow reduction, soil degradation, and technological adaptation (iii). Subsequently, interactions between different subsystems are assessed as important factors in processes of change (iv–vi). These interlinkages are often underestimated in commodity-oriented research. I conclude by examining four socio-economic trends: access to land, transfer of technology, internal differentiation of the peasantry, and the development of a national market (vii–x).

(i) The expansion of pasture is seen by Stonich (1991) and Howard (1989) as the main cause of marginalization of subsistence producers and maize production. In an analysis of census data Stonich (1991) concludes that land in the south of Honduras was reallocated from forest, fallow, and food crops to the production of export crops and livestock. She concludes that the expansion of cattle ranching displaced small producers to marginal highlands, with environmental destruction as a consequence. In this analysis the expansion of cattle ranching by large land holders becomes the simplified representation of change in the agrarian structure. The trend of reduction of forest and fallow area and increase of pasture between 1952 and 1974 did, indeed occur in El Zapote. However, several comments must be added, which change the general picture drawn by these authors.

Firstly, land use patterns (percentage of land used in a specific way) did not change very much between 1965 and 1992, notwithstanding an increase in number of producers. The expansion of pasture between 1952 and 1965 seems to be correlated with a reduction of land under fallow, and not so much with a reduction of food-crop production or forests. To a large extent this land was already used for cattle, in a forest fallow-maize-pasture system, but had never previously been recorded as pasture. In fact, instead of an expansion of area, part of the area was now used for pasture more intensively. Maize and fodder production became separated in the statistics.

Secondly, although the trend in El Zapote has been the concentration of cattle in the hands of fewer people, many small producers with a few head of cattle continued to exist. Sometimes, there is a tendency to redefine agrarian class relations by conceptualizing them in terms of the dominance of cattle and pasture. Certainly, cattle are an important source of revenue

for part of the local elite in El Zapote. But at present they are not expanding cattle keeping as they used to in the past. Furthermore, many small
producers continue to keep cattle and occupy a large percentage of the
area registered as pastures.

Thirdly, pasture for pack animals is important. Coffee has been crucial
for the creation of a class of middle producers in El Zapote. Hence,
pasture expansion cannot simply be related to the development of monopoly capitalism (Howard), with the Junker road of capitalist development,
because it is at the same time related to a process of internal differentiation
of the peasantry.

Fourthly, cattle ranching has been described as extensive because of
low productivity and low labour input per hectare (Howard, 1989). In this
sense cattle have acquired a negative connotation. A more intensive use,
be it grazing or maize cultivation, would intensify soil degradation. The
trend in El Zapote is for pasture to be overstocked (too many animals per
hectare). Producers attest that they would like to intensify production but
are constrained by grass production.

(ii) The expansion of coffee is sometimes explained as a consequence of
earlier cattle expansion. Baumeister (1990) argues that the expansion of
coffee has its origins in the displacement of poor peasants from the
more fertile lands; on hillsides these peasants entered in coffee production which, on slopes, has advantages over staple grains. The available
statistical sources confirm the process of expansion of coffee production.
This expansion accelerated in the 1970s and 1980s, although its founda-
tions were laid down decades earlier. The expansion of coffee started
before the construction of the road to El Zapote and before any institutional development in the coffee sector had taken place. Hence factors
other than infrastructure and institutional development must have been
crucial for coffee expansion. The expansion of coffee, however, did not
lead to the type of producer as conceived by Baumeister. I have argued,
that coffee cultivation is an activity undertaken to generate higher income
and accumulate capital, and which requires some initial capital and labour
resources (Jansen, 1993). Remarkably, the percentage of coffee producers
in El Zapote has declined from 57 per cent of all producers in 1952 to
46 per cent in 1992. With the increasing importance of coffee as cash crop
fewer producers seem to be able to have coffee in their production system.
The trend is towards an internal differentiation of the peasantry in which
coffee plays a pivotal role for accumulating wealth and capital.

(iii) Is technological change a logical and gradual response to population growth, which aims to maintain yields notwithstanding increased
land pressure, shorter fallow periods and soil degradation? This adapted

version of the Boserup model points to some phenomena, but fails to interpret the social construction behind them (Netting, 1993). The population of El Zapote grew from 1228 in 1926, to 2465 in 1950, to 5653 in 1988, and led to almost a same percentage of growth of the number of producers. The reduction of fallow area and fallow period is a very remarkable feature. This is probably linked to a general reduction of mean farm size from 13.6 ha in 1952 to 8.2 ha in 1974, and to 4.0 ha in 1992.

The increase in producers and changed production circumstances might have been a factor that stimulated change, but was not the factor that determined its nature. The new technologies in maize systems have been presented above as responses to a crisis in the production of staple grains. These changes are partly generated endogenously (for example the changes in spacing) and are partly the outcome of an introduction of technology developed exogenously. The use of herbicides, fertilizers, and new cultivars is conditioned by the general emergence of an agriculture oriented on chemical inputs in Honduras. An intensification in use of chemicals was furthermore conditioned by new opportunities offered by an emerging coffee-based agricultural system.

Hence, the notion of the 'natural' character of population growth and its effects should be replaced by a notion of the 'socially constructed' character of population growth (Durham, 1979) in which demographic behaviour is not only a cause but also an outcome of the organization of production. Moreover, the increase in population has not led to an adequate technological change that could accommodate the additional people in agricultural production. Some people are fleeing to the north of Honduras, and one cannot assume that population growth will automatically lead to more producers (made possible by technological change), as out-migration or retreat from agriculture may occur.

Fallow reduction is not a gradual process reflecting the slight increase in pressure on land every year. At various times land pressure was released because large tracts of land became available to villagers. Furthermore, some producers explicitly maintain their fallow period by reducing access of others to their land, and other producers re-introduce a fallow period.

Finally, population growth has increased overall pressure on resources, but the process of dividing the available means of production over the producers has not been equitable. A notion of unequal access to resources is important. Struggles for access influence the nature of production systems. For example, the disappearance of free lands has been related to a drastic decline in the percentage of producers with cattle. The growing importance of coffee as source of income does not mean that people participate in it equally.

(iv) The number of different crops cultivated by a single producer has declined. This is related to the process of social differentiation. The poorest sector has fallen back completely on small plots of maize, and some banana plantings. They claim to have increasing difficulties in sowing beans, planting tule, sowing secondary crops and so on (owing to decreasing access to or availability of land, seed, and labour). The richest sector concentrates on coffee, while the middle sector continues, more than the others, to cultivate a wider variety of crops.

(v) The relation between maize and pigs illustrates how subsystems have feedback relations. In the past maize was used to fatten pigs, which were sold to Salvadorean merchants. On one side, according to producers productivity in maize cultivation has decreased (return to labour) and thus, the costs of pig fattening, in the perspective of producers, have increased. On the other side, the war with El Salvador in 1969 disturbed existing market relations, and later the prohibition of free-ranging pigs jeopardized pig keeping. Pig fattening had been a source for the poor to start some wealth accumulation: to construct their first house or plant a small coffee grove. Pigs were a form of saving, of putting the harvested maize on an account near the house and in a form which could be capitalized through a international trade network. This strategy has been curtailed.

(vi) The interaction between coffee and tule has different modes. Tule is an extractive resource but cannot be placed in the 'traditional', outside capitalist market relations. The growth of a national market for sleeping mats turned tule into a major cash crop during the course of this century. Many producers use coffee revenue to establish tule fields. Hence, coffee can stimulate tule production. Conversely, tule production can support intensification in coffee cultivation when its revenue is used to hire labour or buy inputs. But other producers reduce their activities and hive off tule production, because it takes too much time and diverts them from their principal activity: coffee. While tule cultivation is a complementary activity for the middle strata of producers, it has become inaccessible for the poor and undesired by the richer coffee producers. Most tule fields are far away from the village (2 hours' walk; in the past the fields were near the village) and people without pack animals experience difficulties in transporting the crop.

(vii) Land availability influences which crops and technologies are chosen. This availability has changed remarkably this century. Howard (1989) points to the land value that was historically low and has increased only very recently. This, however, does not mean that processes of unequal land distribution are new. Present-day differential access between producers has its origins in land grabbing, enclosures, and unequal title distribution that started in the last decades of the 19th century.

(viii) The 'availability' of external technology made certain types of technological change possible. Technical change hinges on communication between producers, and much less on project intervention or extension by commercial trading houses. Only the expansion of fertilizer use in maize and the use of new cultivars (maize and coffee) is clearly an effect of development intervention. However, the diffusion of cultivars has not been a simple 'trickle down' process, as producers have selected and adapted technologies and practices to their own production scheme and biophysical conditions.

(ix) Another condition for change was the ongoing internal differentiation of producers. The expansion of coffee demanded an increase in available labour force, especially for picking. The labour force used in El Zapote is almost completely local. Producers use family networks and debt bondage on a small scale to secure sufficient labour. The availability of a cheap labour pool of people without coffee is a condition for the expansion process. It is hard to imagine, for example, producing today's amount of coffee in the situation that prevailed 50 years ago. There would have been an enormous shortage of labour.

(x) The development of a national economy and trade networks has been important; for example, for tule production to expand after demand for sleeping mats increased. Trading not only stimulated crop production but also constrained it, for example a decline of sugar and rice cultivation (although the diminishing rice cultivation is also related to soil degradation).

6 CONCLUSIONS

The description of technological change in El Zapote illustrates the weakness of a dualist view of Honduran agriculture. I have analysed important technological changes in 'subsistence' farming and maize production: many different forms of maize cultivation have evolved and co-exist.

Producers' responses to ecological constraints, to technological change and to socio-economic transformation have generated a wide diversity of production practices. My view about the origins of this diversity differs slightly from authors who have stimulated the discussion about this issue (Long and van der Ploeg, 1994). I attribute less importance to different calculations by producers in the same structural circumstances. In this chapter two types of argument have been presented to illustrate some other mechanisms behind diversity in El Zapote.

The first argument appraises the responses of producers to climatic factors and soil conditions. Small differences in rainfall and temperature and in soil conditions generate big differences in production conditions for different producers. An encompassing analysis of diversity in agricultural production cannot limit itself, therefore, to social dynamics alone (Benton, 1994) and should not throw ecological and demographic variables out of the analysis.

The second argument stresses the importance of understanding of change, because producers cannot all be assumed to be located in similar structural circumstances. Instead, it is imperative to locate producer behaviour in contextual mechanisms that provoke changes. I have argued that these mechanisms cannot be reduced to one single determining process (as the Boserup thesis does, or the monopoly capital argument) but that various processes interact. These are partly organizational-technical (for example, changing ecology, the importance of linkages between subsystems of coffee and pack animals, tule and coffee, pigs and maize, and so on), and partly socio-economic (for example, the changing structural relations of production, demography, and market development and forms of capital accumulation). Hence, a wide variety of different relations and trends shape different production circumstances for each producer. Add, furthermore, the wide variety in biophysical production circumstances and more contingent factors such as household composition, and one gets a picture of the origins of diversity.

Note

1. I am grateful to Sarah Howard, Esther Roquas, and Jan de Groot for their valuable comments, and to the European Commission and the Netherlands Foundation for the Advancement of Tropical Research (WOTRO) for their financial support.

Bibliography

Baumeister, E. (1990) 'El café en Honduras', *Revista Centroamericana de Economía*, Vol. 11 (33), pp. 33–78.

Benton, Ted (1994) 'Biology and Social Theory in the Environmental Debate', in T. Benton and M. Redclift (eds), *Social Theory and the Global Environment* (London: Routledge).

Brouwer, R. & K. Jansen (1989) 'Critical Introductory Notes on Farming Systems Research in Developing Third World Agriculture', *Systems Practice*, Vol. 2(4), pp. 379–95.

Brush, Stephen B. and B.L. Turner II (1987) *Comparative Farming Systems* (New York: The Guilford Press).

DGECH (Dirección General de Estadísticas y Censos de Honduras) (1954) *Censo Agropecuario 1952*, San Salvador.

DGECH (1968) *Segundo Censo Agropecuario Nacional, 1965–1966*, Tegucigalpa.

DGECH (1978) *Tercer Censo Nacional Agropecuario 1974*, Tegucigalpa.

Durham, William H. (1979) *Scarcity and Survival in Central America. Ecological Origins of the Soccer War* (Standford: Standford U.P).

Faber, Daniel (1993) *Environment under Fire. Imperialism and the Ecological Crisis in Central America* (New York: Monthly Review Press).

Galvez, G., M. Colindres, T.M. Gonzalez and J.C. Castaldi (1990) *Honduras: caracterización de los productores de granos básicos*, Colección: temas de seguridad alimentaria no. 7, CADESCA, Panamá.

Howard-Borjas, Patricia (1989) *Implicaciones de la expansión ganadera en la población, el empleo y la alimentación. Alternativas de política a la actual crisis*, Documento de trabajo, SECPLAN/OIT/-FNUAP, Tegucigalpa.

Jansen, Kees (1993) 'Café y Formas de Producción en Honduras', *Revista Centroamericana de Economía*, Vol. 14(41), pp. 58–96.

Jansen, Kees (1995a) 'Ecological Degradation in the Production of Food and Export Crops in North-West Honduras', in M. Mörner and M. Rosendahl (eds) *Threatened Peoples and Environments in the Americas*, Institute of Latin American Studies, University of Stockholm.

Jansen, Kees (1995b) 'The Art of Burning and the Politics of Indigenous Agricultural Knowledge', in Agrarian Questions Organising Committee (ed.), *Agrarian Questions: the Politics of Farming anno 1995: proceedings*, Wageningen Agricultural University, Wageningen, pp. 676–708.

Long, Norman and Jan Douwe van der Ploeg (1994) 'Heterogeneity, actor and structure: towards a reconstitution of the concept of structure', in Booth, D. (ed.), *Rethinking Social Development. Theory, Research and Practice* (Harlow Essex: Longman).

Netting, Robert McC. (1993) *Smallholders, Householders. Farm Families and the Ecology of Intensive, Sustainable Agriculture* (Stanford: Stanford U.P.).

Ploeg, Jan Douwe van der (1990) *Labor, Markets, and Agricultural Production* (Boulder: Westview).

Roquas, Esther (1994) *Las Petateras producen más que artesanía. La economía del tule y del petate*, Documentos de Trabajo No. 8, POSCAE-UNAH, Tegucigalpa.

SECPLAN (Honduras, Secretaría de Planificación, Coordinación y Presupuesto) (1994) *IV Censo Nacional Agropecuario 1993* Graficentro Editores, Tegucigalpa, Honduras.

SECPLAN/DESFIL/USAID, (1990) *Perfil Ambiental de Honduras 1989*, English Summary, Tegucigalpa.

Stonich, Susan C. (1991) 'The Political Economy of Environmental Destruction: Food Security in Southern Honduras', in S. Whiteford and A.E. Ferguson (eds), *Harvest of Want. Hunger and Food Security in Central America and Mexico* (Boulder: Westview Press).

Zúniga Andrade, Edgardo (1990) *Las modalidades de la lluvia en Honduras* (Tegucigalpa: Editorial Guaymuras).

Part III

Natural Resource Reserves and Protected Areas

8 Livelihoods, Land Rights and Sustainable Development in Nicaragua's Bosawas Reserve

Sarah M. Howard[1]

1 INTRODUCTION

The Bosawas National Natural Resource Reserve, designated in 1991, straddles the Department of Jinotega and the North Atlantic Autonomous Region (RAAN), extending from the central mountains to the foothills of the Caribbean lowlands (Figure 7.1) At 8000 square kilometres, it is the largest area of protected humid tropical forest in Central America (SIMAS-CICUTEC, 1995, p. 9). Bosawas is home to the majority of Nicaragua's remaining Mayangna Indians (between 8000 and 10 000), around 4000 Miskitu Indians, and some 30 000 Mestizos (CEPAD, 1992, p. 1; 1993, p. 10; Valenzuela, 1993, p. 14). Mestizo numbers have been increasing since the end of the recent civil war in 1990, through in-migration from other parts of Nicaragua.

This paper examines conflicts over land and resources between Mestizos and Mayangna, tensions between peasant and indigenous livelihoods and conservation, and differences between Mayangna and Mestizo environmental perceptions and land use practices. It evaluates the possibilities for promoting sustainable resource use in Bosawas, within the broader economic and political context. The study is based on interviews and questionnaire surveys conducted in January and February 1995, with 37 respondents from five Mayangna villages on the Bocay River from Tunawalan to Wina and 42 Mestizos, mainly living near Tunawalan (Figure 8.1).

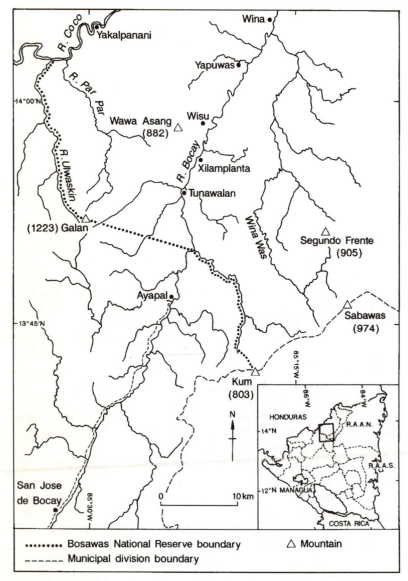

Figure 8.1

2   THE DYNAMICS OF LAND INVASION AND DEFORESTATION

The causes of tropical deforestation include expansion of commercial and peasant agriculture, fuelwood gathering, logging, road building and industrialization. These direct causes are underpinned by population growth, unsustainable agricultural practices, unequal land tenure and government policies (Repetto and Gillis, 1988; Barraclough and Ghimire, 1990; Colchester, 1993a, Utting, 1993). Although all of these factors have contributed to deforestation and dispossession of indigenous peoples in Nicaragua, the main cause has been export agriculture, especially cattle-ranching since the 1950s. Large commercial operators with superior access to credit and the legal system displaced peasants from western Nicaragua eastwards into forested areas in search of land. Property systems, whereby land values and tenure rights accrue through forest clearance, have also encouraged deforestation. During the Somoza dictatorship, settlers received land titles for clearance of so-called national lands (much of it in fact indigenous land) through spontaneous settlement or official colonization schemes (Taylor, 1969). The exhaustion of fragile soils in marginal areas of central and eastern Nicaragua drives settlers further into forested areas. Many sell their cleared land to ranchers. Through land clearance and sale, some peasants accumulate capital and eventually become ranchers themselves. The conversion of forest to pasture also occurs by large operators lending land to peasants for cultivation (Barraclough and Ghimire, 1990; Utting, 1993; CEPAD, 1994). The incorporation of peasants and Indians into the cash economy and population growth exacerbate deforestation (Utting, 1993).

Forest clearance for agriculture slowed during the civil war of the 1980s, but resumed after the war ended. Despite the Sandinista government's agrarian reform, a large proportion of the rural population remained landless. The return of ex-combatants and civilians intensified demands for land and brought renewed agricultural activity and expansion of the agricultural frontier (Cupples, 1992; Ortega, 1992). The electoral defeat of the Sandinistas brought to power a new regime with neo-liberal economic policies, reasserting previous unequal land tenure patterns. Small farmers lost land when it was returned to former owners. Others lost land deposited as security under the new credit laws when they failed to repay loans (Augusto Sandino Foundation/CEPAD, 1994). Structural adjustment policies, adopted in response to Nicaragua's economic crisis under the aegis of the World Bank and International Monetary Fund, have reduced credit for small farmers, promoted agro-exports and encouraged cattle-raising as the only credit-worthy option. Deforestation for timber proceeds apace by companies set up by former state employees (who lost their jobs in government spending cuts) and illegal operators, while economic problems continue to restrict funding of environmental protection.

3   MESTIZO ORIGINS AND LAND TENURE IN THE BOCAY AREA OF BOSAWAS

Almost half of respondents were born in the Department of Jinotega (including one born in Tunawalan and two in the municipal division of Bocay), almost a third were born in the adjacent department of Nueva Segovia and only three were born in western Nicaragua. Almost two-thirds of respondents were living in the Bosawas area prior to the war, the earliest since the late 1960s. However, the remainder arrived for the first time after 1990, and further arrivals are expected as other family members join existing settlers and former residents return. Besides those born locally or who arrived as children, a fifth of respondents left their original locality because they did not have their own land, almost a quarter left because there was insufficient land and just over a tenth left because their land was poor.

The first reason reflects the fact that a fifth of respondents were landless in their place of origin, while around a third worked their parents' land. Insufficient land among adult households probably arose from sub-division of family land, exacerbated by large family sizes. However, three respondents who owned land individually in their place of origin also complained of insufficient land, each holding less than 50 *manzanas* (one manzana is equal to 0.7 hectare). Poor land quality and declining crop yields, which caused some respondents who had their own land to leave, reflects the role of unsustainable agriculture in migration to forest areas. That two respondents retain land in their place of origin suggests opportunistic rather than impelled migration.

Most respondents are, on the basis of Agriculture Ministry categories, small land-holders. Almost half hold up to 50 manzanas, about a quarter hold 50 to 100 manzanas and the remainder hold 100 to 300 manzanas (MIDINRA's (nd)). None could be classified as large landowners (with over 500 *manzanas*). Besides those working land bought by their parents, almost half of respondents farm land purchased from other Mestizos, around a tenth bought land from local Mayangna, one demarcated land in the Somoza era and two are borrowing land. In all, almost half of respondents work land originally bought from the Mayangna.

4   MESTIZO VIEWS OF LAND AND RESOURCE CONFLICTS

The main concern among the Mestizos is over the designation of Bosawas. Only one respondent favours Bosawas (for environmental reasons), with the remainder almost equally divided between those opposed and those who are unclear about Bosawas. Resentment of Bosawas by Mestizos partly arises from the lack of consultation with local inhabitants over its designation. Mestizos also fear being evicted from Bosawas. Having lost crops and houses when they were evacuated during the war they do not want further upheaval. Mestizos suspect the government of plans to remove them, leaving only indigenous people. Although plans to relocate Mestizos were overtaken by the return of former residents, discussions of policy have been confined to meetings held in the main town between project staff and community leaders, allowing uncertainties to persist among the rural population. Fears of relocation were heightened after an environmental non-governmental organization (NGO) placed signs on the borders of Bosawas stating that the area was part of a project for development of the indigenous communities of Jinotega and the Coco River. The Mestizos interpreted this as a threat to their right of occupancy and removed the sign on their side of the river, along with replacement signs installed later by the Mayangna. Although the incident over the signs catalyzed contention between the Mestizos and the Mayangna, and a fifth of Mestizos criticize the Mayangna for trying to reclaim land which they had sold in the past, most Mestizos blame uncertainties over land rights in Bosawas on the government.

Mestizos also resent being forbidden to cut wood within Bosawas for sale in Ayapal and San Jose de Bocay (Figure 8.1), and having to ask permission to cut trees on their land for other purposes. They view instructions to restrict clearance of virgin forest as thwarting their ambitions for economic progress through agriculture. Prohibition of land sales within Bosawas (intended to prevent further in-migration) is also unpopular. Mestizos believe that other Nicaraguans are entitled to settle in Bosawas, so they continue to subdivide and sell land.

5   MAYANGNA VIEWS OF LAND AND RESOURCE CONFLICTS

Almost all Mayangna respondents were born on the Bocay River. According to older Mayangna, before the Mestizos arrived, the Mayangna were few and families used land in the area according to need. There was no concept of a bounded Mayangna territory. Unlike the indigenous people of the RAAN, whose land rights are recognised under the Autonomy Statute and whose political interests are represented by a multiethnic Regional Council, the indigenous peoples of Jinotega have no institutionalized political representation or legislative recognition of their land rights. To defend their interests, the Mayangna communities of the Bocay River with Miskitu communities from the upper Coco River have formed the Association for the Development of the Sumu and Miskitu Peoples of Jinotega (ADEPSOMISUJIN). During 1994, with the support of a environmental NGO and the leaders of ADEPSOMISUJIN, community leaders delimited the area which they claim as Mayangna territory, including land occupied by Mestizos. Mayangna leaders do not seek to evict existing settlers, but they want recognition of a continuous Mayangna territory to prevent further incursion.

Conflict over land with Mestizos is the greatest concern among the Mayangna. They are worried by the continual influx of Mestizos and the subdivision and sale of land by established settlers. Although the Mayangna sold land to Mestizos in the past, they firmly state that they no longer do so, because this would leave insufficient land for future generations. Moreover, they accuse Mestizos of taking larger areas than they actually bought. In places such as Wina and Tunawalan the Mestizo population now far outnumbers the Mayangna population. The Mayangna feel they are being invaded and fear that soon there will be insufficient land for themselves and their children.

Fear of land scarcity in the 1970s led the Mayangna to delimit private land parcels to prevent land being taken by Mestizos. Almost half of respondents work land which has been privately demarcated, and seven respondents (two extended families) work land which was privately purchased. However, I was assured that these private plots will be incorporated into communally-titled areas. The private holdings are the subject of disputes between individual Mayangna and Mestizos. Almost half of Mayangna respondents complain that Mestizos are clearing and working land within their parcels. Other conflicts are likely to arise between former landowners who have not yet returned and Mayangna families who are now working their land.

Although Mayangna leaders, particularly those from the RAAN, have condemned the designation of Bosawas reserve as a violation of indigenous territorial sovereignty, and three respondents oppose Bosawas for similar reasons, the remainder are equally divided between those who are unclear about Bosawas and those in favour. The high percentage in favour of Bosawas is probably because, as a result of support for Mayangna territorial claims by NGOs, the Mayangna have come to view Bosawas as syn-onymous with defence of their land and resources against Mestizo invasion.

6 MESTIZO AND MAYANGNA IDEOLOGIES

Mestizos and Mayangna view themselves and each other in terms of polarized stereotypes – the lazy Mayangna versus the productive Mestizos, the environmental guardians versus the destroyers of the forest – which underpin their claims to land and resources and their attitudes to the Bosawas reserve. The Mayangna justify their territorial claims on the grounds of their history of occupation of the area, their rights to territory as indigenous people and on the basis of being the natural custodians of the forest. They contrast themselves to the destructive and irresponsible Mestizos.

The Mestizos claim the right to work within Bosawas on the grounds that they are Nicaraguan citizens. Whereas some claim rights through purchase, others justify entitlement to land through the hard work of clearance. Mestizos portray themselves as hard-working, supporting the rest of the country through their activities, while they regard the indigenous people as lazy. The designation of Bosawas as a national reserve may encourage in-migration, since it is perceived to be national land and therefore free for the taking. Many Mestizo respondents have a strong attachment to their locality, and a few stated that they are prepared to fight for their land if necessary.

Despite opposing the Bosawas reserve, some Mestizos are aware that deforestation can disrupt the hydrological cycle and destroy vital resources, having seen this happen elsewhere in Nicaragua. However, respondents argue that they need to clear virgin forest, because it is not possible to obtain good harvests by successive cultivation of the same plot. While almost all Mayangna and Mestizo respondents think it necessary to look after the forest, almost half of the Mayangna compared with a third of Mestizos wish to protect the forest for environmental reasons, while Mestizos are much more likely than Mayangna to see forest protection in terms of safeguarding resources. Whereas over half of Mestizos would favour the arrival of a lumber company because it would generate jobs and bring roads, almost two thirds of Mayangna think that a lumber company would destroy the forest. Almost three quarters of Mayangna think that the forest could one day disappear, compared with less than half of Mestizos.

Four-fifths of Mayangna are concerned about the loss of forest fauna, which they blame on the destruction of their habitat and over-hunting by the Mestizos. In contrast, only a fifth of Mestizos think that forest animals are disappearing. Most think that they are increasing and they complain that animals eat their crops and livestock. The Mayangna see themselves as part of the environment, and regard the forest in its entirety as a communal good. Mestizos perceive the forest in terms of potential capital and resources, owned as part of individual plots. They tend not to perceive the need to retain forest in its own right and many regard the idea of protecting animal habitats as absurd.

It can be argued that the stereotypical polarization of Mestizos and Mayangna attitudes is an over-simplification. Nevertheless, there are differences in the way in which each group perceives the environment, and a greater concern over environmental destruction on the part of the Mayangna. This may reflect the long history of residence of the Mayangna in the area, their more recent incorporation into the cash economy and greater exposure to the work of environmental NGOs.

7 MESTIZO AND MAYANGNA LAND AND RESOURCE USE PRACTICES

Perceptions of the resource base, and accompanying practices of land use, have implications for efforts to promote sustainable development within Bosawas. The main difference between the land use practices of the Mestizos and those of the Mayangna is the ownership of cattle and planting of cattle pasture. Whereas only two Mayangna have cattle pasture, and none have cattle, two thirds of Mestizos have pasture and half have cattle. Mayangna customarily graze cattle on small areas of unfenced, natural pasture whereas Mestizos tend to create large areas of sown and fenced pasture. However, most of the Mestizos with pasture only have up to ten manzanas and are small cattle-owners with less than ten cows. Only three have ten cows or more. Cattle-raising provides economic security in view of uncertain markets for crops and transport difficulties (Utting, 1993, p. 19). The permanent removal of forest for the creation of extensive areas of cattle pasture could well become an increasing threat to Bosawas.

Some Mayangna practice long fallows of up to 20 years, while some Mestizos leave no fallow or replace crops with pasture. However, the average fallow period is the same for both groups, around two years. Differences in areas planted to major crops also varies little, although the Mayangna plant smaller areas of corn, while sowing larger areas of root crops (such as cassava and *quequisque*) and bananas and plantains. Although root crops are nutritionally demanding, they provide longer ground cover than other crops, as do bananas and plantains. This helps to reduce soil erosion. The Mayangna on average also clear smaller areas of forest annually for crops (1.9 *manzana* compared with 2.7 *manzanas*).

In terms of use of forest resources, around one third of Mestizo households claim to hunt forest animals, compared with just over a half of Mayangna households. Mayangna are also more likely to gather forest plants than Mestizos. Neither Mayangna nor Mestizos admit to selling lumber, firewood or forest animals, probably because of restrictions placed by Bosawas. Some of the Mayangna say that they occasionally used to sell forest animals but they no longer do so because of Bosawas.

Some aspects of Mayangna resource use are potentially more sustainable than those of Mestizos, since they tend not to create pasture, clear smaller areas of forest, depend more closely on the forest habitat, and plant crops which provide longer ground cover. The fact that Mestizos hold large forest areas within their individual lots is worrying, since this is likely to be sold for clearance or converted to pasture. However, differences between the two groups may decrease over time. The evacuation of Mayangna communities during the war, pressure on land, and the fact that the land had almost ten years' rest in their absence, seems to have undermined traditional long-fallow practices, while greater market integration has led Mayangna to increase areas of crops (SELAS-CADESCA, 1994, p. 18). Moreover, some Mayangna had cattle before the war, and others aspire to do so. Inter-marriage between Mayangna and Mestizos, which already occurs, may also erode differences between Mayangna and Mestizo land use practices.

8   PROSPECTS FOR SUSTAINABLE DEVELOPMENT

Policy makers have recognised the need to give local people a stake in the management of reserves such as Bosawas, to encourage their protection and to reduce the costs of policing reserves. Management strategies increasingly focus on attempts to preserve core areas of undisturbed forest, while permitting human activities within the surrounding buffer areas (Ghimire, 1994, p. 199; FUNDESCA-GRET, 1994; Utting, 1994). Such strategies underpin management plans for Bosawas. The survival of the Bosawas forest, and the long-term welfare of its inhabitants, depends upon control of in-migration; stabilization of the existing population and their participation in the defence and sustainable management of the reserve. The principal strategy adopted by the Bosawas project, under the aegis of the Nicaraguan Ministry for Natural Resources and the Environment

(MARENA) and with support from the Swedish and German governments, has been selecting and training local people to form a network of local forest guards. Forest guards (most of whom are volunteers) monitor reserve invasion and illegal extraction of lumber and try to persuade local agriculturalists to minimize forest felling and to stop selling land. In the Bocay area the two guards are indigenous people.

Since 1992, various NGOs and international donor agencies have begun to develop initiatives in the Bosawas area, including the European Community, Oxfam UK, Danish Development Assistance (DANIDA) and USAID. Projects funded by Oxfam and DANIDA are being executed by Nicaraguan NGOs such as the Alexander Von Humboldt Centre and the Foundation for Autonomy and Development of the Atlantic Coast (FADCANIC). USAID funds work by the North American organization The Nature Conservancy (TNC). The German Society for Technical Co operation (GTZ) is also influential in policy development for Bosawas. The strategies of these organizations for stabilizing the local population within the buffer zone include addressing Mayangna and Mestizo land claims, intensifying production and improving land management through promotion of natural composts, field crop diversification, tree crops, and silvo-pastoralism. Forest regeneration, community forest management and development of non-timber products will be encouraged to give people a stake in forest defence.

These initiatives face a number of challenges. Projects will have to provide alternatives to would-be colonists and offer concrete incentives to local agriculturlists to modify their land use practices. At the time of this research, forest guards were able to offer only advice about what not to do, and in the Bocay area they had stopped working after receiving death threats from Mestizo settlers. Greater dialogue is required with the Mestizos and greater efforts to integrate them into the Bosawas project. Although Mestizo and Mayangna attitudes and resource use practices differ, and Mayangna may be more open to alternative resource use strategies than Mestizos because they have not yet focused on cattle-raising, is also worth recognizing the similarities between the two groups. Furthermore, although Mayangna land claims need to be prioritized in view of their present lack of land security, the Mestizos are the largest group in Bosawas and the survival of this reserve depends upon their co-operation.

It is important that the introduction of sustainable alternatives accounts for local economic aspirations, and is achieved through genuine dialogue rather than imposition of outsiders' knowledge (Utting, 1994, pp. 232–3). This requires visits to local communities as well as discussions with community leaders. The dilemma facing project promoters is that such a dialogue will be a medium to long-term process, while there is an immediate need to find solutions in order to gain local co-operation and control deforestation.

Successful management of Bosawas also requires close co-ordination between the diverse state institutions whose policies affect the reserve, particularly with regard to land titling. The issue of land titling is contentious. Whereas attempts to promote sustainable forest use depend upon security of tenure, titles enable landowners to gain credit for ranching, and can encourage land speculation. Nevertheless, the development of procedures for titling indigenous lands is vital. Although the Nicaraguan Institute for Agrarian Reform (INRA) is working towards such measures, in the mean time, indigenous people either must wait until appropriate legislation is developed, risking loss of further land, or claim land rights within the existing framework, which undermines customary land use practices and obliges them to relinquish claims to territorial sovereignty (Jarquin, 1993).

Progress has been made towards the sustainable development of Bosawas in the last year, bringing closer co-ordination between governmental and non-governmental organisations and local representatives, and consolidation of a policy for reserve management. However, attempts to defend Bosawas must still contend with national political and economic problems.

9   CONCLUSIONS

The situation of land claims within Bosawas is complex. Policy makers face the challenge of safeguarding indigenous land rights while gaining the co-operation of the Mestizo inhabitants, enabling participation of local people in the sustainable management of Bosawas. Although the potential exists for promoting sustainable resource use among Mestizos and Mayangna, success will depend upon resolution of wider political and economic problems, particularly the unequal access to land throughout Nicaragua. This in turn will be influenced by decisions taken by Nicaragua's international donors and creditors.

Note

1. I would like to thank the many people in Nicaragua, including personnel of governmental and non-governmental organizations, who supported my work and gave up their time to be interviewed. I would also like to thank the

140   *Nicaragua's Bosawas Reserve*

Economic and Social Research Council for funding the research through award number R000221294. I am very grateful to Hilary Foxwell for drawing the map.

Bibliography

Augusto C. Sandino Foundation and the Council of Evangelical Churches of Nicaragua (CEPAD) (1994) *Environment and Development – Dilemmas, Challenges and Achievements of Nicaraguan NGOs.*

Barraclough, S. and K. Ghimire (1990) 'The Social Dynamics of Deforestation' in *Developing Countries Principles, Issues and Priorities*, UNRISD Discussion Paper Number 16.

CEPAD (1992) 'Bosawas – Will Protecting Bosawas Protect the Sumu?', *The CEPAD Report*, May–June 1992, pp. 1–9.

CEPAD (1993) 'The Battle for Bosawas – Will Anything be Left for Posterity?', *The CEPAD Report*, July–August 1993, pp. 9–12.

Colchester, Marcus (1993a) 'Colonizing the Rainforest: The Agents and Causes of Deforestation, in Colchester, M. and Lohmann, L. (eds), *The Struggle for Land and the Fate of the Forests*, World Rainforest Movement, Penang.

Colchester, Marcus (1993b) 'Forest Peoples and Sustainability' in Colchester, M. and Lohmann, L. (eds) *The Struggle For Land and the Fate of the Forests*, World Rainforest Movement, Penang.

Cupples, J. (1992) 'Ownership and Privatization in Post-Revolutionary Nicaragua', *Bulletin of Latin American Research*, Vol. 11, No. 3, pp. 296–306.

FUNDESCA-GRET, (1994) *Programa de Desarrollo Sostenible en Zonas de Frontera Agricola en Centro America, Fase Preparatoria, Informe Final* (unpublished report).

Ghimire, Krishna (1994) 'Parks and People – Madagascar and Thailand', *Development and Change*, Vol. 25, No. 1, pp. 195–229.

Jarquin Chavarria, L. (1993) *Diagnostico Legal Acerca de la Titulacion de Tierras de las Comunidades Indigenas Ubicadas en la Reserva Bosawas* (unpublished).

MIDINRA, (nd), *La Ley de Reforma Agraria* (unpublished).

Ortega, M. (1992) *Diagnostico Bosawas: Siuna, Rosita, Bonanza*, Centro de Investigacion ITZANI (unpublished).

Presidencia de la Republica (1991) 'Declaracion de la Reserva Nacional de Recursos Naturales Bosawas, Decreto No. 44–91', *La Gaceta, Diario Oficial*, 5 Noviembre, Ano XCV, No. 208.

Repetto, T. and Gillis, M. (1988) *Public Policies and the Misuse of Forest Resources* (Cambridge: World Resources Institute and Cambridge University Press).

SELAS-CADESCA (1994) 'Programa de Frontera Agricola, Diagnostica de Bosawas', Nicaragua, Documento de Trabajo, No. 3 (unpublished).

SIMAS-CICUTEC (1995) *Bosawas – Frontera Agricola, Frontera Institucional?* (Report Commissioned for The Nature Conservancy).

Taylor, R. (1969) *Agricultural Settlement and Development in Eastern Nicaragua*, Madison, University of Wisconsin, Institute of Agricultural Economics, Research Paper Number 33.

Utting, P. (1993) *Trees, People and Power* (London: Earthscan).

Utting, P. (1994) 'The Social and Political Dimensions of Environmental Protection in Central America', *Development and Change*, Vol. 25, No. 1 pp. 231–59.

Valenzuela, M. (1993). 'The Bosawas Nature Reserve and the Indigenous Communities: An Uncertain Future', *Barricada Internacional*, Vol. XIII, No. 362, pp. 14–16.

9 Alternative Approaches and Problems in Protected Area Management and Forest Conservation in Honduras

Michael Richards[1]

1 INTRODUCTION

Honduras, with a forested area of between 4.5 and 5 million ha (40–45 per cent of the total land area), is probably the Central American country with most forest left. Just over half is estimated to be broadleaf, and the rest coniferous. As in other countries in the region, deforestation has been very rapid over the last 30 years; Utting (1993) estimates that Honduras lost about 25 per cent of its forests between 1964 and 1986; while USAID (in Schreuder, 1992) puts the loss at a third between 1964 and 1990. Deforestation of broadleaf forest was in a range of 65 000 ha to 80 000 ha (3–4 per cent p.a.) – at this rate all the broadleaf forest would disappear within 20 years. This reflects a government policy of frontier expansion, as shown by agrarian reform, credit, road construction and tenure policies (Brockett, 1990). Treating the more remote areas as an-escape valve for the landless has evaded the politically difficult process of redistributive land reform. Structural adjustment and recent neo-liberal economic reforms have if anything deepened this trend.

During the 1980s, the top-down response to deforestation in the region was to adopt a 'fences and fines' approach in which authorities tried to keep people out of an ever increasing area under protection. In Central America, the number of national parks and reserves increased from 30 in 1970 to more than 300 in 1987, covering about one sixth of the total land area (Utting, 1993). In Honduras, 104 protected areas (PAs) were created between 1980 and 1994.

However deforestation meanwhile increased sharply to about 2 per cent per annum (Utting, 1993). Rigid conservationist approaches, associated with government agencies in charge of PAs, attempted (where there were sufficient resources) to deny local people the use values of the forest, alienating them from conservation objectives. Also, for under-resourced forest or national parks authorities, it soon became apparent that it was physically impossible to stop encroachment by loggers, ranchers and colonist farmers. Such factors have led donors to promote a more participatory and largely non-government organization (NGO) led approach to forest conservation. These efforts have tended to focus on the stabilisation of farming systems in colonization areas or buffer zones, and environmental education.

This chapter focuses on two major PAs in Honduras, the Rio Platano Biosphere Reserve (RPBR) in the tropical moist forest area of eastern Honduras, and La Tigra National Park, a cloud forest in the mountains of central southern Honduras (see Figure 9.1). It looks at how government agencies and NGOs have responded to the major problems in the two PAs, and how either a more regulatory or participatory/market-based strategy has been adopted so far. The viability of these two broad approaches to forest conservation, and of the possibilities along the continuum between them, is considered. This leads to a discussion of the need for a more balanced approach, and the constraints against this in Honduras at present. Most of the data in this chapter were collected by the author in 1994 from discussions with local people, NGOs and national agencies.

2   THE RIO PLATANO BIOSPHERE RESERVE

The Rio Platano Biosphere Reserve (RPBR), was the first (in 1980) Biosphere Reserve in Central America, and with an area of 525 000 ha, easily Honduras' largest protected area. Most of this is lowland tropical rainforest, although there is some mixed broadleaf and pine forest in the southern buffer zone. It has a high species diversity in several large untouched areas, including many endangered species. It forms a human habitat for three indigenous groups: the Miskito Indians, the (Afro-Caribbean) Garifuna and the endangered Pech. The forested central mountainous area is the main source of, or feeds into, the principal watersheds of the area of Honduras known as Mosquitia: the Platano, Patuca, Paulaya and Wampu rivers. The area also includes valuable ecotourist resources such as white water rapids, extensive Mayan ruins and sandy beaches on the north coast.

The 207 000 ha nuclear zone of the RPBR is fortunately protected on most sides by natural barriers. However access has proved relatively easy from the south. It is estimated that at least 25 per cent of the southern area

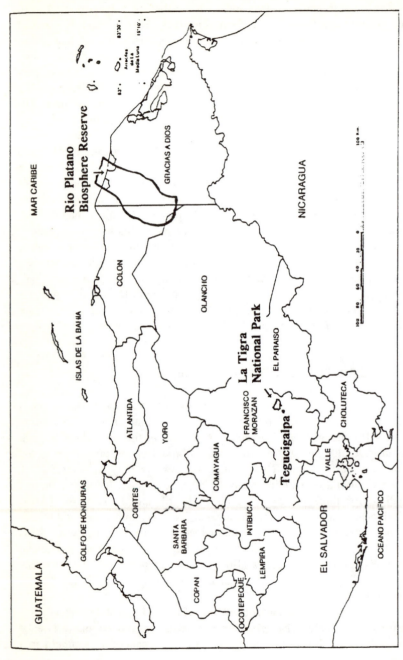

Figure 9.1

was deforested or badly degraded between 1980 to 1992, with considerable incursion into the nuclear zone (GFA, 1992). The population, livelihood systems, associated environmental problems, and other characteristics of the PA are summarised in Table 9.1.

3 LA TIGRA NATIONAL PARK

La Tigra National Park was also made a PA in 1980. Its 25 800 ha is mainly montane cloud forest: 80 per cent of the 7600 ha nuclear zone was over 1800 m, most of it steeply sloping (50–75 per cent). The forest is deciduous above 1500 m, with a mixture of broad-leaved and pine forest on the lower slopes. Of 31 mammal species in the park, six were considered to be in danger of extinction, including mountain lion, margay, ocelot, and coataquil (Romero and Martinez, 1992).

Unlike the RPBR, La Tigra's forests have a high realisable market value, mainly due to its proximity to the capital, Tegucigalpa. La Tigra supplied about 40 per cent of Tegucigalpa's water (Romero & Martinez, 1992). Flooding was also a major threat to the city – severe flooding occurred about once every two years. These factors indicate the very high real or opportunity cost of forest conversion. However, high market values for cash crops and natural resources have resulted in major deforestation pressures. Some of these pressures and other characteristics of La Tigra are summarised in Table 9.1.

4 THE MAIN PROBLEMS FACING FOREST CONSERVATION IN THE PAs

There was considerable similarity in terms of the problems faced in the two PAs. These included land tenure issues, legal and institutional conflicts, the human presence in the nuclear zone, and the environmental consequences of unsustainable land use practices, as indicated in Table 9.1.

Land tenure

Land tenure in both areas is best summed-up as chaotic. In both PAs, large areas of land were effectively under private and municipal ownership, in spite of PA legislation designating it all as national forest land. For example in La Tigra, about 30 per cent of the nuclear zone was held by

Table 9.1   Main characteristics of the protected areas

	Rio Platano Biosphere reserve	La Tigra National Park
Forest type	Lowland tropical rainforest in north, mixed pine/broadleaf forest in south	Montane and cloud forest, pine/deciduous mixed forest below 1500 m
Area	525 000 ha with nuclear zone of 207 000 ha	25 820 ha with nuclear zone of 7571 ha
Population/ indigenous people	30 000 mostly Ladinos; 4500 Miskito, smaller numbers of Garifuna and Pech (in danger of extinction); about 3700, including some military, in nuclear zone	3800 of which about 1000 live in nuclear zone – including two major political families; multinational banana company.
Livelihood basis	*Ladinos*: cattle, coffee, timber, game, food/cash crops from slash and burn; Miskitos: lobster diving, destructive harvesting of NTFPs, esp. when bans on lobster extraction, slash and burn Garifuna and Pech: less destructive extractivism, especially fishing; slash and burn inc. fallow enrichment.	Middle-class or wealthier farmers: coffee, cattle, timber. Campesinos: these activities, but greater emphasis on slash and burn; methods, intensive vegetable and flower growing, firewood collection and sale, etc.
Main environmental problems	ranching, chemicals in nuclear zone; over-exploitation of fauna in north; erosion of extractivist basis of indigenous groups; contaminated and silting rivers, reduced navigability; illegal logging, ranching in south; climate changes reported by local people, eg less rainfall, longer dry season, winds; forest fires in south.	20% of nuclear zone converted for coffee, cardamon agrochemicals in nuclear zone affecting water sources; soil erosion due to steep hillside cropping, especially vegetables and coffee, increasing flood risk and damage in Tegucigalpa; illegal timber extraction; fires due to arson and pasture maintenance
Land tenure situation	Indigenous land rights mapped but not recognized; *Ladinos* claimed land with guns and by crudely marking boundaries; mixture of national forest land, 'municipal' and 'private' land with conflict between these	30% of land in nuclear zone in private hands, especially political and commercial elites, including family of an ex-President and a multinational banana company; mixture of

Table 9.1 Continued

	Rio Platano Biosphere reserve	La Tigra National Park
	tenure types. PA legislation ineffective; Further confusion from 1992 legislation.	'private' and 'municipal' land in buffer zones, conflicting with PA legislation
Balance of market and non-market values	High non-market values, especially watershed protection and biodiversity; low market values except timber in southern buffer zone; high existence value associated with three indigenous groups; unrealised eco-tourism values.	High market values, especially water (supplies 40% of the capital's supply, also vital for farming), timber, eco-tourism (unrealised); coffee and vegetable farming on cleared land; also high non-market values: watershed protection, biodiversity
Government Agencies	COHDEFOR (Dept. PAs & Wildlife) took over from Sec. of Natural Resources in 1991, but rarely entered the PA due to resources and inherited unpopularity	COHDEFOR from 1991, usually only enters the Park with armed escort; Secretariat of Natural Resources; SANAA – Water Authority
Main NGOs	Honduran Ecology Assoc. (AHE) to 1990, World Neighbours in south; MOPAWI and BAYAN in north working with indigenous groups	Honduran Ecology Assoc. to 1992 World Neighbours (1987–91) Friends of La Tigra

political and business elites, including two prominent political families and a multinational banana company. Quasi-legal ownership rights have been established through repeated land transactions and the crude physical demarcation methods of *Ladino*[2] colonists. Most 'private' land was really only *de facto* possession. While in some areas, such 'owners' were not worried by lack of title or documents, in others conflicting claims were common. Lawyers were often brought in to provide pseudo-legal documentation.

In the RPBR, indigenous land rights were mapped (through Cultural Survival's support) but not recognised. The drastic erosion of the area occupied by the Pech, in the face of *Ladino* guns, has meant that this indigenous group is in danger of extinction – with only an estimated 1800 left (Salaverri, 1992).

Particularly in the RPBR, different interpretations of the 1992 Agricultural Modernization Law, which represents a neo-liberal attempt to free up the land market, has caused immense confusion and considerable speculation. This has been largely due to a clause that *de facto* possession of forest land can be converted into full title after three years. While this was not supposed to national forest land in buffer zones, actual land transactions have reflected the view that it applied to all forest land. In remote colonization areas, what is more important is not what the law says, but how it is interpreted and acted upon.

Human presence in the nuclear zones

A second common problem in the two PAs was the significant human population in the nuclear zone, many of who were engaged in degrading natural resource practices (Table 9.1). Especially in La Tigra these were politically influential, and have prevented any effective control of land use practices. In the case of the RPBR, a US $9 million German grant has been held up since 1990 due to an unmet condition that the nuclear zone population be relocated or, in the case of the majority who entered illegally since 1980, expelled.

Market and development pressures

As indicated in Table 9.1, a major contrast between the PAs was the difference in the realizable market value of the forest. The proximity of La Tigra to the main national market explains the prevalence, in both the nuclear and buffer zones, of capital-intensive cash crops like hybrid coffee, cardamon, vegetables, flowers and other crops. These were often grown on steep hillsides, causing soil erosion and chemical contamination of watercourses – DDT levels have been recorded 400 per cent above permitted levels (SANAA, undated). Timber and firewood were also under high pressure; much of this was extracted at night to avoid problems with authorities. Land in the buffer zone was also in demand from wealthy Tegucigalpan commuters.

In the case of the RPBR, market pressures were less important than subsistence/development pressures, although *Ladino* land uses, including ranching and coffee farming, were mainly market-based. There was also over-exploitation of fauna (iguana, armadillos, tapir, deer and a range of other game) and non-timber forest products like palm hearts due to the demands of both market (mainly by *Ladinos*) and subsistence (mainly by Miskitos). *Ladinos* have formed hunting clubs and use 'trained' hunting dogs.

5 THE RESPONSE: INSTITUTIONS, STRATEGIES AND PROBLEMS

State agencies

In both areas, the Secretariat of Natural Resources (SRN) was responsible for PA management until 1991, although its presence in the RPBR was minimal. In 1991, official management of the PAs was reluctantly handed over to the newly created Department of Protected Areas and Wildlife (DAPVS) of the Honduran Corporation of Forest Development (COHDEFOR). The other main state agency involved was the National Water Authority (SANAA), which had armed guards around the main micro-watersheds of La Tigra.

COHDEFOR's impact on the colonization processes in the two PAs has been minimal. This was firstly due to a lack of resources – it inherited only the responsibility and not the resources from SRN. Also COHDE-FOR became very unpopular in the two areas for repeatedly giving permission to wealthy outsiders to extract timber, while prohibiting local people from using the forest, or even clearing low bush/fallow areas for subsistence farming. For example, during the 1980s a sawmill owned by a prominent politician was allowed to operate on the edge of the RPBR buffer zone, while a small farmer cooperative to extract resin from the nearby pine forests was shut down (Salaverri, 1992). COHDEFOR was also suspected by local people to be in collusion with illegal timber and fauna extraction by the military. This made it very difficult for COHDEFOR to initiate community based approaches, when it decided to move away from a more regulatory approach.

A second issue has been institutional conflict. SRN has always resented handing over responsibility to COHDEFOR, and has made the task of the latter particularly difficult, especially in La Tigra. This came to a head when the COHDEFOR Park Director was physically intimidated by local SRN staff when trying to develop a community eco-tourism project in 1991. The latter threatened the income stream of SRN forest guards, who supplemented their meagre salaries by doubling up as tourist guides. Following fierce criticism of COHDEFOR by SRN, an NGO called AMITIGRA (Friends of La Tigra) was formed, which assumed management control in 1993 through a Presidential decree. Many saw this NGO as 'the government in disguise' – most of AMITIGRA's founding members had close affiliations with the SRN. The President of Honduras was the main landowner in the nuclear zone and a former Minister of SRN. The other main government agencies felt that AMITIGRA lacked the technical capacity to manage the Park.

Institutional overlap was also a major problem in La Tigra: Romero and Martinez (1992) reported that 20 state agencies and 18 NGOs were involved in some way in La Tigra in 1990, and that 'functions and responsibilities of each are so unclear that it is common to find two institutions doing the same thing' (p. 46). For example there was a clear overlap in fire protection, encroachment and land use control between COHDEFOR, SRN, SANAA and AMITIGRA. Such problems have contributed to the failure to implement any of the five detailed management plans that have been prepared for La Tigra.

In contrast, in the RPBR such overlap problems have not occurred – due to the virtual absence of state agencies. However in both areas, the Armed Forces have been brought in at intervals to 'defend the environment'. Local people reported widespread indiscipline by soldiers (such as hunting game when off-duty); they were clearly very frightened and alienated by the military.

Another major problem in both areas was the conflict between local government and COHDEFOR. This was a question of the perceived hierarchy of legislation. Municipalities have interpreted the 1992 Municipalities Law as conferring on them management responsibility for buffer zone 'municipal' land. The author came across many cases where Municipalities had given permission to farmers to clear forest for farming, or granted private land rights on purchase of municipal land – in one case to a whole community in La Tigra. More recent PA legislation has given authority to COHDEFOR over all buffer zone land, but this appears to have had little effect so far – at least partly due to its weak and unpopular presence in buffer zone areas.

Non-Government Organisations

In the RPBR, in the absence of an effective state presence, the NGO community has largely taken over PA Management with international donor support. The NGO approach evolved from a more protectionist or regulatory strategy to a participatory or community orientated approach. The Honduran Ecology Association (AHE), which did most to alert the national and international communities to the gravity of the problems in the 1980s, worked in both PAs for several years until it became discredited in an administrative mismanagement problem in 1992. In the RPBR, it at first used a policing approach with the support of the Honduran Armed Forces, emphasizing prohibition of felling, hunting, etc. AHE became so unpopular that it was unable to implement the environmental education and other participatory elements of its programme.

The World Neighbours (WN) programme in the southern buffer zone of RPBR was regarded by the Canadian World Wide Fund for Nature (WWF) as a model project. However for the first two years (from 1990), WN found it very difficult to get acceptance in the area. This was because they were introduced by AHE, and initially some of the same staff were employed, so local people thought WN would adopt the same regulatory approach. The situation improved with the recruitment of local people in 1992.

WN has made good progress in the introduction of sustainable farming methods in the buffer zones of both PAs,[3] slowing down migratory farming pressures. By 1994, WN estimated that some 600 farmers in the RPBR were using green manure cover crops and minimum tillage (in-row ploughing only) techniques. The use of green manures like velvet beans (*Mucuna* spp.) interplanted with maize was a traditional practice of small farmers in northern Honduras. This technology is now reported to have spread throughout Central America (Pasos *et al.*, 1994) and has enabled farmers to crop semi-continuously the same piece of land with a minimal decline in fertility. This is due to at least three major benefits from green manures and minimum or zero tillage: soil enrichment from biomass build-up (nutrient recycling) and nitrogen fixation; suppression of weed growth, creating a labour-efficient alternative to burning; and protection of the soil during the dry season from heat and erosion (Buckles, 1994).

An important recent action of WN has been promotion of the grass-roots Agroforestry Cooperative of the Rio Platano, which is keen to develop sustainable forest management activities, like collection of pine resin. Successful community-based programmes have also been developed by other NGOs like Mopawi and Bayan working with indigenous groups in the northern buffer zone.

6 DISCUSSION: VIABILITY OF REGULATORY AND MARKET-BASED APPROACHES

Conservation efforts in PAs by government agencies and/or NGOs can vary along a continuum between more regulatory ('fences and fines') approaches, and market-based strategies which depend on developing adequate market incentives for the reconciliation of development and conservation objectives – thus inducing participation of local people. Some of the characteristics of the 'regulatory approach' would be as follows:

- excluding people from the nuclear zone, involving relocation/ compensation the people who lived there before it was declared a PA, and ejection of illegal settlers;
- strict control of land uses in the buffer zones;
- control by the authorities of a further influx of people into the buffer zones.

The main problems of this approach are the cost, and the hardening of attitudes of the excluded stakeholders against conservation. The market approach depends on giving local people a greater protective interest in forest conservation, mainly through:

- increasing direct forest use values to local people, for example through promoting sustainable forest management, community-based eco-tourism, drinking water projects, etc.; such activities, which can create positive incentives linking conservation and development objectives, were identified as the vital missing dimension in a number of protected area management projects, reviewed in a major study by Wells and Brandon (1993);
- greater security of tenure to those living in the nuclear and buffer zones, and providing them with effective extension and training support, so that they can form a 'human wall'.

There are also several limitations to the market-based approach. These include the problem that the value of many forest products is often not high or consistent enough to make forest management more attractive than other land uses – for example the generally depressed and fluctuating prices of natural rubber and Brazil nuts have raised important economic question marks about the viability of the Brazilian Extractive Reserves. However timber, eco-tourism and some of the high-value low volume NTFPs hold untapped economic potential. A second limitation is the power of local commercial elites to suppress market development by forest communities – the market approach may require more political will than a regulatory approach. A third limitation can be the high institutional demands in terms of support services for market-based development.

The stabilization of farming systems in buffer zone areas could (and should) be part of both of these approaches, although given the necessity for a highly participatory approach to technology development in forest margin areas, it is more likely to be effective when linked to the market approach. As WN's work has shown, this has the potential to reduce pressure on core conservation areas, but care is needed to ensure that this takes place on soils which are ultimately appropriate for farming – according to SECPLAN *et al.* (1990), 80 per cent of Honduras' land is classified as basically of forest vocation.

Viability in the Rio Platano Biosphere Reserve

In the Rio Platano Biosphere Reserve, arguably the main limitation on a more regulatory approach is the size and remoteness of the area – the level of resources to effectively exclude people completely from the core zone, and control the land use practices of a large and diverse buffer zone population would be relatively high. The success, albeit on a relatively small scale, of the NGOs in promoting a participatory approach contrasts with the earlier policing efforts. However it can be argued that NGOs have been relatively slow to develop incentives that link conservation and development objectives – as for example in forest management. This is partly because market-based incentives are inevitably less strong in a remote area, but untapped opportunities clearly exist for community-based eco-tourism (especially in the north) and pine resin extraction in the southern buffer zone. In favour of this approach there have been a number of positive experiences in Honduras of community-based natural forest management – for example among the pit-sawing cooperatives of the north coast (Richards, 1993), and pine resin extractors in central Honduras (Stanley, 1991).

However it would be unwise to rely solely on the market-based, participatory approach. Certain strong and unambiguous regulatory actions are required by government, for example ejection of illegal squatters in the nuclear zone,[4] and strict control of land use practices of the 40 or so bona fide families which could remain. Standing up to the powerful vested interest groups who have-moved into the area, like the ex-army officers turned ranchers, is one of the most urgent and politically complex actions required. Also a state presence in the buffer zone is urgently necessary to ensure correct implementation of recent land and PA legislation.

Viability in La Tigra National Park

Unlike RPBR, the theoretical viability of both the regulatory and market approaches in La Tigra are much higher. The regulatory approach is much more viable in view of the access and size of the area. While there are at least 50 forest guards between the various agencies in the dry season (the number varies with the perceived fire risk) with direct or indirect policing duties, logistical back-up (vehicles, fire-fighting equipment, etc.) has remained weak. They have been unable to control the actions of the powerful interest groups present. For example, a SANAA (nd.) report listed five people who committed extensive damage in 1992: these included a colonel, a cousin of the President, and a foreigner.

The original idea was to make La Tigra a model area for a participatory approach to forest conservation. However problems of institutional infighting, the unpopularity of state agencies, land tenure chaos, unregulated market-driven land uses and political elites, have ensured that the high forest direct use values (especially from water and eco-tourism) have not been converted into conservation incentives. Income, from water revenue, should have been available to finance these, but SANAA has used this to subsidise its national operations. Thus the political/institutional, rather than economic, viability of the more participatory approach is the real problem in La Tigra. In view of this, and the huge values at stake – water is regarded as life or death for Tegucigalpa – the feeling of leading environmentalists is that time has run out for the participatory approach after a 'decade of failed social experimentation' (Jorge Betancourt, ex-President of AHE, personal communication).

7 CONCLUSION

This analysis shows that, in order to counteract the problems identified, an array of actions are needed which fall between the two extremes – the market and regulatory approaches – and that a balanced approach is required that regulates the harmful impact of human interventions, especially in the core areas, and encourages positive attitudes in surrounding buffer zones.

At present, the prospects for a more effective regulatory role of government are not good – there is insufficient political will to expel or control the powerful elites involved, especially in La Tigra; anti-state attitudes are high; and Honduras is too poor (as well as lacking political will) to dedicate sufficient resources to PAs. COHDEFOR has experienced difficulties persuading local people of its new PA management role. Although not without its problems, the recent creation of the National Parks Service in Costa Rica could be an important experience for Honduras to consider.

Ironically, while market values should mean that the market/participatory approach is more viable in La Tigra, in fact the prospects seem better in the RPBR. The positive experiences of the NGOs in the RPBR need to be built on. However the long-term success of such approaches is linked to regularization of tenure and effective implementation of land and PA legislation – therefore more effective government presence in the PAs is essential.

There are, or should be, obvious complementarities between government and NGOs in developing and harmonising more regulatory and participatory approaches. For example, policies which do not treat the forest as an expandable frontier, and creation of a stable institutional environment are essential for reducing risks, increasing the time-horizon of the participants, and encouraging internalization of externality-type forest benefits in the market/participatory approach. Also control of in-migration could be needed to prevent forest margin areas becoming 'attraction zones' if the market or participatory approach becomes 'too successful'.

This potential synergy of roles is confused when a pseudo-NGO like AMITIGRA is created, and the military is involved in policing. The desire of the Honduran military to find a *raison d'être* as defenders of the environment is particularly worrying. Thus the holistic approach required must call forth, as well as political will, much greater sensitivity to the interplay of regulatory and participatory approaches, institutional coordination, and external (donor) support than hitherto evidenced in Honduras.

Notes

1. The author gratefully acknowledges the support and interest of all those who cooperated in the research, but especially COHDEFOR, Miriam Dagen King (World Neighbours), Torsten Kowal (CONSEFORH), Peter Utting (UNRISD) and Andy Thorpe (Portsmouth University) for commenting on the draft. The research was funded by a grant from the European Commission (DG8) to the ODI for forest margins research. However the author takes sole responsibility for any errors.
2. Mixed-blood descendants of Indians, Europeans and Africans, that form the majority of Honduras' population.
3. WN has also had a stabilising impact on buffer zone communities from four years' work (1977–81) with farming systems on the eastern edge of La Tigra's nuclear zone.
4. Failure to do this has encouraged illegal encroachment into other PAs.

Bibliography

Brockett C. (1990) *Land, Power and Poverty: Agrarian Transformation and Political Conflict in Central America* (Boston: Unwin/Hyman).
Buckles D. (1994) *Velvetbean: a 'new' plant with a history*. CIMMYT Economics Programme Internal Document, Mexico.

COHDEFOR/FAO (1989) *Estadisticas Forestales*, Tegucigalpa, Honduras.

GFA (1992) *Proyecto Manejo de la Reserva de la Biosfera del Rio Platano*, Feasibility Study Final Report for COHDEFOR and KFW (German Bank of Reconstruction), Tegucigalpa, Honduras.

Richards M. (1993) 'Lessons for Participatory Natural Forest Management in Latin America: Case Studies from Honduras, Mexico and Peru', *Journal of World Forest Resource Management* 7, pp. 1–25.

Romero R. and Martinez F. (1992) 'In Honduras: Water for a Thirsty City', in V. Barzetti and Y. Rovinski, *Towards a Green Central America. Integrating Conservation and Development* (Connecticut: Kumarian Press).

Salaverri J. (1992) 'La Situación Actual de la Reserva', in V. Murphy, *La Reserva de la Biosfera del Río Platano: Herencia de Nuestro Pasado*, Paseo Pantera, Tegucigalpa, Honduras.

SANAA (nd) *Problemas*, Mimeo, Tegucigalpa, Honduras.

Schreuder G. (1992) *Environmental analysis of the impact of the new Honduran Agricultural Modernization Law with respect to the Forestry Sector*, Consultancy Report for Inter-American Development Bank, Tegucigalpa, Honduras.

Schuerholtz G. (1991) *La Tigra, Interim and Final Reports for CIDA*, Tegucigalpa, Honduras.

SECPLAN, DESFIL and USAID (1990) *Perfil Ambiental de Honduras*, Funded by US Agency for International Development, Tegucigalpa, Honduras.

Stanley D. (1991) 'Communal Forest Management: the Honduran Resin Tappers', *Development and Change*, Vol. 22, pp. 757–79.

Utting P. (1993) *Trees, People and Power. Social dimensions of deforestation and forest protection in Central America* (London: Earthscan).

Walker I., Suazo J., Thomas A., and Jean-Pois H. (1993) *El Impacto del Ajuste Estructural Sobre el Medio Ambiente en Honduras*, Central American Postgraduate Economics Dept., National Autonomous University of Honduras, Tegucigalpa, Honduras.

Walker I. (nd) *Concluding thoughts from CEASPA study on the environmental impacts of structural adjustment in the forestry sector in Honduras*, Mimeo, Tegucigalpa, Honduras.

Wells M. & Brandon K. (1993) *People and Parks. Linking Protected Area Management with Local Communities*, World Bank/WWF/USAID, Washington DC.

Part IV

Agrarian Policies For Sustainable Land Use

10 Land Reform and Resource Management within Agrarian Production Cooperatives in Honduras

Ruerd Ruben and Marrit van den Berg

1 INTRODUCTION

Agrarian production cooperatives (APCs) established during the Land Reform process in Honduras face serious problems with respect to income generation, labour mobilization and land use. Therefore, pressure to proceed with parcellation of these cooperative enterprises is imminent. Major arguments for parcellation refer to the incentives required to improve the efficiency of resource allocation and, to a lesser extent, for the creation of favourable prospects for sustainable land use.

The main objective of this chapter is to analyse the management regimes that determine input allocation within the cooperative context. The analysis is based on a critical review of the frequently heard hypothesis that cooperative property tends to favour resource depletion. Instead, it will be argued that Honduran Land Reform cooperatives are well prepared to cope with interdependence in contract choice, and their parcellation may even reduce prospects for efficient resource management. APC consolidation requires internal informational procedures to be reinforced in order to reduce uncertainty within a well defined bargaining framework. Our methodology strongly relies on *contract-choice* theory as an institutional economic approach for the analysis of organizations, giving priority to mechanisms of coordination of transactions among participating agents (Putterman, 1981; Douma and Schreuder, 1991). Much attention is given to the effects of organizational and allocation rules on enterprise performance, as well as the incentives that are implicitly incorporated into exchange rules. Contrary to the traditional paradigm of the 'tragedy of the commons', collective action could be a feasible device to guarantee viable incentives for maintaining of the natural resource base (Bromley, 1992). Therefore, economic efficiency and sustainability are enforced simultaneously within a contract-choice framework, thus creating new prospects for the consolidation of land reform cooperatives.

The paper is structured along the following lines. In Section 2, different theoretical perspectives on resource use in a common property context are compared with respect to their implications for cooperative organization. Conceptual issues for the analysis of potential (dis)advantages of collectively owned and managed resources are highlighted, and different managerial factors and decision making procedures that influence resource allocation within APCs are analysed. This theoretical contemplation is followed by an empirical appraisal of intra-firm efficiency differences in resource management in Section 3. This section is based on field research made in Comayagua in 1994/95 among 172 collective and individual producers. The market policies actually implemented to reinforce APC land use management (titling, pricing) are discussed in Section 4, and compared with contract choice alternatives that could improve resource allocation and sustainable land use within a cooperative context.

2   COMMON PROPERTY

Management of common property resources

Management systems for common property resources represent an important issue of the current policy debate on cooperative performance (Platteau, 1992). Common property is defined as private property for a well-defined group, with bounded membership and internal allocational mechanisms for individual usufruct rights, while behaviour of all group members is subject to a set of generally accepted rules. Two different approaches can be distinguished for the analysis of resource management within common property regime: *property rights* approach, and *contract choice* theory.

Property rights

The property rights approach is based on the enforcement through markets as a prerequisite for investment and relies basically on external incentives for land conservation. Two different paradigms regarding common property prevail: (i) the *prisoner's dilemma* model of strategic choice, (ii) Olson's *'logic of collective action'*. These models were developed within the framework of game-theory, but rely on a somewhat different specification of principal-agent relations, and the related policy consequences for institutional change differ widely.

The prisoner's dilemma game is based on completely independent decision making by individuals. Although cooperation would yield higher pay-off for all agents, each agent individually faces an incentive to free ride. This outcome is modified, however, if agents are permitted to negotiate with each other on allocational rules, or if the game is played repeatedly, taking learning processes into account. Penalties could be used to increase incentives to cooperate and reduce pay-off for individual free riding.

The argument supplied by Olsen (1971) in *The Logic of Collective Action* can be seen as a further extension to the prisoner's dilemma. Rational individual choice will promote free-riding behaviour, unless individual contributions to group objectives can be enforced through selective coercion. External agents are required to regulate access and resource allocation according to commonly agreed criteria. Wade (1987) criticizes this proposition because potential availability of net collective benefits will be an equally important incentive for voluntary and self-enforcing co-operation. Differences in collective action could be attributed to factors like group size and homogeneity, costs of exclusion, extent of mutual obligations, discipline and relative immunity from state interference.

The major shortcoming of the property rights approach refers to the fact that it disregards risk and selective market failure. Resource depletion due to overuse (e.g. exceeding carrying capacity) is likely to arise from three related phenomena: (i) a *boundary problem* caused by demographic pressure and rivalry in consumption among participating agents; (ii) an *internal assurance* (or *coordination*) *problem*, due to uncertainty about the expected behaviour of other agents within the group; and (iii) an *external insurance problem*, due to the occurrence of multiple failures on land and capital markets. Property rights theory can only address the first two problems that limit incentives for investment in the maintenance of the soil quality of common property resources. Low willingness (propensity) to invest is then related to *uncertainty* about expected income due to limited knowledge about the behaviour of other participants and problems of guaranteeing direct income flows from private investment, causing high *time preference* or low *discount rate* for future benefits and resulting in an intratemporal allocation of labour biased in favour of productive activities and neglecting conservation activities.

Contract choice

Contract choice theory offers alternative internal incentive schemes to improve employment and investment (Hayami and Otsuka, 1993). As the private economic revenue of each agent depends on how intensively all other agents participating in the same resource base use the resources (interlinked decisions), the expected utility of each agent is contingent on information about the actions of all other agents. Potential advantages of coordinated decision making on resource use and allocation are clarified within a framework of strategic interdependence among agents under conditions of uncertainty, which warrant a stable Nash conjecture.[1] Whether such an outcome occurs, depends on the following conditions:

(a) The presence of *assurance* mechanisms. These are required, as transboundary effects of individual management decisions cause optimum efforts and investments to be dependent on expected actions of other agents. Nonseparable costs or outputs are therefore an incentive for coordination and enhance scale economies. Sufficient access to information on prevailing market conditions is the major condition to be satisfied to permit members to objectively evaluate opportunity costs of cooperation as a substitute for missing markets.

(b) Defined *monitoring* systems that permit enforcement of labour efforts and control for free-riding behaviour. As individual efforts are conditional upon expectations about the probable (cooperative or defective) behaviour of other agents, clear mechanisms for internal control and sanctions are required.

(c) *Coordination* of activities among individual members. This is required to enforce strategic interdependence. Mutual dependency relies on reciprocal mechanisms for conditional commitment of individual participating agents (Runge, 1984). Recognition of interdependence requires that collective benefits of joint action should be at least sufficient to compensate for transaction costs of cooperation (Wade, 1987), as all decision makers need a private economic reason for collective participation.

(d) *Services* supplied by the cooperative enterprise to individual members, introducing an element of *interlinked transactions* into the member-firm relationship. In terms of contract-choice theory, these interlinked transactions and long-term contracts based on reputation act as a guarantee for contract enforcement (Hayami and Otsuka, 1993). Reduced behavioural uncertainty in turn enables a lower discount rate for future benefits and permits contributions of APC membership to future welfare to be valued.

(e) *Participation* of all members in the election and control of APC management. This is a condition *sine qua non* for internal democracy and accountability of leadership. These mechanisms should guarantee the long-term prospects of the enterprise and permit multi-period assurance mechanisms to be effective. Intertemporal welfare redistribution can only be achieved when individuals are willing to accept suboptimal outcomes in the short run because they are assured that others will compensate them over time.

These five principles offer a consistent framework for the appraisal of rationality for the emergence and evolution of stable common property institutions that are capable of satisfying members' expectations and offer assurance against free-riding behaviour and resource depletion. The contract-choice approach also guarantees that production and distribution aspects are directly related, and therefore a *bargaining solution* is required to satisfy efficiency and equity considerations. Moreover, cooperation prevents the consumption of common property resources by one agent imposing negative externalities on others, thus precluding for excessive aggregate use and promoting investment for resource maintenance (Ostrom and Gardner, 1993).

Resource management in Honduran land reform cooperatives

Cooperative enterprises in Honduras can best be characterized as a system of '*collectively linked private parcels*' (Carter *et al.*, 1993), indicating that they operate as a 'coalition of participants' with simultaneous decisions on land use, factor allocation and internal services. The structure of the cooperative system is composed of three related components: (i) cooperative enterprise, (ii) family home plots, and (iii) households (Fekete *et al.*, 1976). The cooperative enterprise represents the business organization in charge of resource mobilization and allotment of home plots, sometimes also possessing formal property rights. Family home plots are individually cultivated parcels of land that guarantee mainly food security and part of market supply, making use of services offered by the cooperative enterprise. Households are engaged in decisions on labour allocation among collectively and individually cultivated parcels, off-farm employment and leisure, and are in charge of decisions about the division of expenditure between consumption and savings. Interactions among these entrepreneurial and organizational dimensions within the cooperative system determine resource allocation and patterns of land use, as well as prospects for sustainability.

Institutional organization of common property regimes can be analysed as a response to local physical, economic or technological conditions. The endogenous character of contract-choice is then considered as an interaction among (i) external physical and socio-economic environmental conditions, (ii) internal allocative and distributive procedures, and (iii) interactions among cooperative members (Oakerson, 1992).

Physical resources and technical attributes

The boundary conditions for APC performance refer to the availability of physical resources (land, labour, capital) and the technical attributes that determine (i) the capacity for joint use (such as distance to markets and the share of available land suitable for cropping activities), (ii) the possibilities for exclusion (that is admission procedures), and (iii) the scale of production.

Agricultural production in most APCs depends mostly on access to land and complementary material inputs. Average availability of land within Honduran APCs (4.8 ha/family) is considered insufficient to sustain the farm-household economy, and joint use is thus enforced for reasons of access to (short-term) credit and material inputs that permit more intensive land use. In water-limited situations, access to irrigation is another incentive for joint use.

Decision making mechanisms

Procedures for decision making on resource use are set within the boundaries of internal operational rules and external conditions. Three different aspects can be distinguished: (i) procedures for planning, (ii) payment rules, and (iii) information on markets.

Most important planning procedures refer to the frequency of production planning and the scheduling of joint marketing operations as factors that reinforce group interest. Planning of labour services necessary for collective production may further enhance coordination. It contributes to a greater transparency of planned operations, although it may also limit the flexibility of factor allocation during time.

Internal rules for the remuneration of labour efforts (wages or piecework), frequency and stability of wage payments, and rules for distribution of profits among the members represent the most important direct incentives for members to participate in collective production.

Information and knowledge about prevailing conditions on factor and commodity markets represents a third dimension of decision making. Knowledge on available options for factor mobilization through land, capital (credit and co-investment) and labour markets, as well as on alternative marketing outlets permits a more thorough appraisal of costs and benefits for internal resource allocation.

Patterns of interaction

Interaction among APC members can be related to five sets of variables: (i) monitoring systems, (ii) assurance mechanisms, (iii) direct coordination among members, (iv) services supplied by the cooperative towards members, and (v) internal governance.

Monitoring of members labour efforts is dependent on the use of supervision mechanisms and the application of sanctions in the case of free-riding behaviour. Both procedures tend to reinforce labour discipline in collective production. Otherwise, most important assurance mechanisms include kinship ties (*parentesco*) and membership of a peasant union, both offering a wider framework to cover behavioural risks.

Coordination among members with respect to soil preparation, input purchase, and pests and weed control for their individual land parcels indicates positive externalities of APC membership. Moreover, services supplied by the cooperative enterprise to individual members, like input supply, transport, marketing or provision of training and welfare are expected to increase prospects for private economic returns.

Finally, different aspects of internal organization reflect the dynamics of membership participation in APC management and the accountability of internal governance. Periodic elections of members of the cooperative board, and internal decentralization through a number of committees are indicators of democratic management and control. Moreover, stability of the composition of the board could guarantee accumulation of experience, while frequent rotation of the board tends to affect continuity of operations.

Performance and results

The performance of APCs can be evaluated with three sets of criteria: (i) efficiency of resource use, (ii) equity of income distribution, and (iii) sustainability of the resource base. In the following section we will present a case study in which we will mainly address efficiency aspects and indicate how these are related to the prospects for sustainable resource management.

3 TECHNICAL AND ALLOCATIVE EFFICIENCY OF APC RESOURCE USE IN HONDURAS

Technology choice and efficiency of resource use within the Honduran cooperative sector can be analysed at two different levels: (i) comparing absolute levels of input use between the reformed and the private sector, and (ii) comparing allocational differences between individually and collectively cultivated parcels of land within the Land Reform cooperatives. In this paper we will follow the second procedure, indicating – whenever possible – implications for sustainable resource use.

An analysis of the efficiency of internal resource allocation permits the importance of relevant managerial factors that determine differences in resource use among collective and individual parcels to be identified. Based on fieldwork in the central department of Comayagua (Van den Berg, 1995), we estimated technical and allocative efficiency scores on the basis of deviations from the OLS estimates of the loglinear Cobb-Douglas production function. Scores for technical inefficiency were determined on the basis of the frontier production function. Allocative inefficiency was determined as the ratio between the marginal value of the produce and the marginal costs of inputs. Correlation analysis was used to select significant managerial and organizational variables that explain differences in cooperative performance, and common factor analysis was applied to derive latent factors (Hair *et al.*, 1992). Our random sample included 172 cases, composed of 27 collective farms and 145 individual members of cooperatives. The field survey refers to the 1994/95 agricultural cycle and included questions on actual input use and farm-gate prices for all agricultural products, and a separate review of current management practices and organizational rules (see section 2 for the general framework). For our field research, the relevant variables were measured with a 5-point Likert scale that ranged from low to high interaction.

Resource use

The whole farm production function was estimated, including separate variables for the type of management (0 for collective organization and 1 for individual organization), and specific crops (proportion of cultivated area under horticultural or perennial crops). Moreover, mechanical traction was specified as the proportion of the cropping area where this input is used.

The results of the OLS estimate are given in Table 10.1. Except for traction, all input coefficients have the expected positive sign. The traction variable includes expenses for transportation, which do not influence the output of production. The tractor dummy better explains the positive result of mechanical soil preparation. The insignificance of the coefficient for fertilizer at a 10 per cent level can be explained

Table 10.1   Coefficients of the production function for 27 APCs in Honduras

Variable	Coefficient	Standard error	Significance
Inputs			
land	0.604661	0.142949	0.0000
labour	0.299390	0.095791	0.0021
seed	0.100559	0.047741	0.0367
fertilizer	0.123517	0.089623	0.1701
herbicides	0.072249	0.027644	0.0098
pesticides	0.101809	0.026344	0.0002
traction	0.047978	0.027743	0.0857
Other variables			
tractorized area	0.549374	0.167475	0.0013
horticultural crops	−1.035911	0.397101	0.0100
perennial crops	1.266152	0.191014	0.0000
management type	0.196151	0.188117	0.2987
Constant	5.564278	0.480513	0.0000

Significance: Adjusted R^2 = 0.84.

by heteroscedasticity. The proportion of the total area under permanent crops has a strong positive impact on revenue, but horticultural crops give the opposite effect; this may be due to the higher risks involved and their sensitivity to adverse weather conditions (1994/95 was a very dry year).

Individual management seems to be technically more efficient than collective management; however this effect is not statistically significant at a 10 per cent level. Apparently, incentive problems hardly exist. Still, management has an indirect influence on the technical efficiency of the farms in the sample. On collective units, statistically significantly more perennial and horticultural crops are cultivated. If neither of the crop variables is incorporated in the function, the management coefficient is statistically significant and has a value of 0.44. This means that when crop choice is considered as a variable dependent on management type, full parcellation would yield a 55 per cent increase in gross monetary output. On the other hand, positive economies of scale are confirmed, summing the coefficients of the essential inputs. The combined results of these contradictory processes indicate that positive economic return to resource sharing can only be expected when a minimum number of agents are willing to cooperate.

Technical efficiency

Technical efficiency scores are defined for both individually and collectively cultivated parcels in order to identify relevant factors that explain factor-neutral efficiency differences. To do this, differences between actual gross monetary revenue and potential revenue according to the frontier production function (corrected for extreme outliers) were ascertained and statistical correlation with the structural characteristics of individual or collective management regimes were determined.

For collective management, only 3 significant variables were selected: (i) collective marketing ($p = 0.39$), (ii) frequency of wage payment ($p = 0.61$), (iii) payment of membership subscriptions ($p = -0.61$). The first is typically an interaction variable, while the other two belong to the category of decision making rules. Collective marketing enhances efficiency, as this collective service increases the benefits of cooperation for the individual members. Frequent payment of wages is an important direct incentive for the peasants. The payment of membership subscriptions decreases efficiency. Peasants that have fulfilled this obligation to the cooperative have more leeway to evade other collective obligations. Factor analysis was performed for the five variables which influence the technical efficiency of individual production. The resulting factors are: (i) labour organization on the collective plot; and (ii) cooperation in individual production. The presence of a collective employment guarantee ($p = -0.15$), frequent meetings ($p = -0.14$) and labour supervision systems ($p = -0.14$) appear to tie labour to the collective plot and lower the flexibility of individual production. Coordination of input purchase ($p = 0.15$) and joint tractor use ($p = 0.15$) improve technical efficiency. This voluntary cooperation points towards positive externalities of APC membership.

Allocative efficiency

Differences in allocative efficiency between collective and individual production were determined through the univariate Mann-Whitney test, which indicated significant variation of the efficiency score.[2] Seed, herbicides and traction demonstrate significant group differences (Table 10.2). The application of seeds and herbicides is allocatively more efficient in individual production, while traction is generally underused under both management regimes.

The use of the labour force is generally efficient on both collectively and individually cultivated plots. This indicates that labour markets function rather well and that the practice of remunerating collective labour with regular advance payments (*planillas*) seems to act as a sufficient incentive to participate widely in collective operations (Ruben and Roebeling, 1995). This is also confirmed by the analysis of the composition of the labour force at individual and collective levels, which indicates the reliance on wage and family labour at individual level permits members to participate in collective work in order to maintain rights to wages and profit sharing (Ruben *et al.*, 1992).

Seed is used extravagantly on collective plots and applied efficiently on individual plots. This reflects the practice that individual peasants rely on their own seeds kept from the latest harvest and hardly buy any additional seed, whereas for collective production most seed is bought at the market. Fertilizer, herbicides and traction are strongly underutilized on both types of plots. This indicates that there is a severe shortage of cash and credit. Pesticides, however, are efficiently applied on average. Apparently, peasants give them high priority.

The correlation matrices of the allocative efficiency scores for collective and individual production with relevant institutional and organizational parameters are presented in Tables 10.3 and 10.4. In total, 46 explanatory variables at individual level and 30 variables at collective level give a

Table 10.2   Confidence intervals for allocative efficiency on 27 APCs in Honduras

Input	Mann–Whitney[*]	90% confidence intervals for the allocative efficiency score[**]		
	Significance	Collective	Individual	Total
labour	0.4154			(–0.80, 0.87)
seed	0.0479	(–3.83,–0.12)	(–0.02, 0.34)	
fertilizer	0.6753			(0.52, 0.81)
herbicides	0.0027	(0.92, 0.99)	(–0.10, 0.85)	(0.04, 0.86)
pesticides	0.7175			(–0.30, 0.80)
traction	0.0021	(1.28, 3.94)	(2.64, 3.55)	

[*]   A significance level below 0.10 means that the hypothesis that there is no difference between collective and individual production is rejected.
[**]   A score of 0 signifies allocative efficiency. Negative values indicate excessive use. Positive indicate that too little of the input is applied.

Table 10.3 Correlation between the allocative efficiency scores* for collective production and production unit characteristics

Variable	Labour	Seed	Fertilizer	Pesticides	Herbicides	Traction
Technical attributes						
distance to the market					.82 (.046)	
age of the cooperative					.89 (.017)	
number of associates					-.76 (.080)	
Internal governance						
participation in the board			.41 (.080)	.41 (.099)		
stability of the board			.40 (.086)	-.46 (.065)		-.49 (.021)
family relationships among member			.51 (.024)	.54 (.027)		.57 (.006)
Assurance mechanisms						
union membership		-.56 (.003)	-.57 (.011)	-.58 (.015)		-.58 (.005)
collective employment guarantee			-.89 (.000)	-.94 (.000)		-.90 (.000)
credit availability					.80 (.059)	
tenancy	-.42 (.026)					
Procedures for planning						
frequency of production planning		-.39 (.044)	-.68 (.001)	-.71 (.001)		-.61 (.003)
division of labour on collective plot		.39 (.047)	-.47 (.042)	.61 (.009)		.43 (.046)
planning of marketing			-.52 (.022)	-.66 (.004)	.80 (.059)	-.53 (.011)
frequency of meetings				-.42 (.091)		-.38 (.082)

Table 10.3 Continued

Variable	Labour	Seed	Fertilizer	Pesticides	Herbicides	Traction
Cooperative services						
emergency help			-.40 (.090)	-.50 (0.46)	.80 (.046)	-.55 (.008)
provision of training						-.42 (.050)
wage payment in case of illness						-.36 (.098)
provision of traction or irrigation					.93 (.008)	
collective marketing			-.43 (.063)	-.43 (0.87)		
coordination of marketing of individual produce					.80 (.057)	
Monitoring mechanisms						
sanctions			-.45 (.054)	-.45 (.971)		-.59 (.004)
supervision			-.47 (.040)	-.55 (.021)		-.62 (.002)
Market information						
knowledge of the capital market			-.50 (.029)	-.50 (.042)		
knowledge of coinversion			-.39 (.098)			
Payment rules						
stability of wages			-.58 (.009)	-.54 (0.26)		-.45 (.034)
frequency of wage payment		-.33 (.088)	-.46 (.047)	-.49 (.048)		-.46 (.032)
daily wages (high) or piece-wages (low)		-.40 (.040)	-.76 (.000)	-.79 (.000)		-.70 (.000)
collective profit capacity			-.56 (.014)	-.49 (.047)		-.54 (.009)
retainment of profits			.49 (.035)	.44 (.077)		.44 (.039)
rules for profit distribution		-.41 (.036)	-.70 (.000)	-.73 (.001)		-.71 (.000)

* The allocative efficiency scores reflect allocative *in*efficiency, so negative correlation with the scores means positive correlation with allocative efficiency.

Table 10.4 Correlation between the allocative efficiency scores* for individual production and production unit characteristics

Variable	Labour	Seed	Fertilizer	Pesticides	Herbicides	Traction
Technical attributes						
fraction of agr. land under arable crops	−.26 (.002)					
fraction of cultivated land in collective use	−.18 (.024)					
number of associates					.42 (.007)	−.16 (.092)
Internal governance						
membership desertion		.17 (.048)				
participation in the board	.23 (.005)	.22 (.007)	−.22 (.020)			
stability of the board					.28 (.076)	
democratic planning	−.17 (.038)					
Assurance mechanisms						
union membership	−.21 (.011)					−.31 (.001)
membership of other organizations			.31 (.001)			
admission of new members			.21 (.026)			
control of finance						
Cooperative services						
emergency help			.16 (.095)	−.28 (.012)		
assistance in house building				−.20 (.091)		
provision of inputs for individual production			−.18 (.006)			
wage payment in case of illness				−.30 (.006)		

Table 10.4 Continued

Variable	Labour	Seed	Fertilizer	Pesticides	Herbicides	Traction
collective processing	.34 (.000)	.22 (.009)				.28 (.003)
collective transport	.16 (.055)					
Coordination among members						
coordination of soil preparation			-.17 (.075)	.22 (.049)		
coordination of purchase of inputs				.23 (.037)		
labour exchange	.21 (.010)		.17 (.075)	-.28 (.012)	.28 (.075)	
cooperation in marketing	.22 (.008)					
planning of marketing				-.26 (.017)		
frequency of meetings			-.22 (.021)			
Monitoring mechanisms						
sanctions			.18 (.069)			
supervision						.21 (.025)
Market information						
knowledge of the land market				-.20 (.072)		
knowledge of the capital market				-.24 (.033)		
knowledge of the product market	.23 (.005)	.22 (.009)			.29 (.075)	
knowledge of coinversion	-2.3 (.004)					
Payment rules						
stability of wages		.20 (.003)				
frequency of wage payment		.33 (.034)				

Table 10.4   Continued

Variable	Labour	Seed	Fertilizer	Pesticides	Herbicides	Traction
retainment of profits	.17 (.043)		.16 (.092)			−.16 (.091)
payment of membership quotas	−.21 (0.11)					
Personal attributes						
share of cooperative wages in income	−.19 (.021)					
labour days on collective plot						
labour days off-farm	.21 (.010)			−.28 (.012)		
availability of assets	−.14 (.097)					
born inside/outside the region		−.15 (.023)				
years of residence in region				−.19 (.019)		
ratio of dependents/non-dependents	−.16 (.050)					

* The allocative efficiency scores reflect allocative inefficiency, so negative correlation with the score means positive correlation with allocative efficiency.

significant correlation. Variables with negative correlation coefficients are positively correlated with allocative efficiency. The variables that are most important for allocative efficiency of each of the production factors were analyzed separately, using factor analysis. For collective production, herbicides were excluded from further analysis, as there were only few cases.

Collective production

The allocative efficiency of *labour use* is only statistically significantly correlated with one variable, namely tenancy. The existence of an official title to the land decreases the uncertainty of expected income. Surprisingly, there are no other statistically significant correlations. Many of the factors investigated were expected to influence the incentives to the members. Apparently, incentive differentials do not influence labour efficiency directly, but operate through the efficiency of the use of the other inputs.

Seed efficiency is strongly influenced by the internal labour organization of the cooperative. Only one factor emerges from factor analysis: clear rules for labour allocation and profit distribution. It involves variables such as guaranteed collective employment, clear rules for profit distribution, and manner of wage payment (frequency, daily wages instead of payment for piece-work), which all enhance certainty of income flows. The frequent planning of production schemes permits timely adjustment of operations. A strict division of labour hampers flexibility.

Efficiency in the use of *fertilizers* is influenced by a large number of variables, that can be grouped into five factors: (i) clear rules for labour allocation and profit distribution; (ii) planning procedures; (iii) internal governance; (iv) market information; and (v) services. The first factor coincides with the enumerated effects on seed efficiency of payment and allocation rules. Family relationships appear to hinder compliance with these rules. Assistance of the cooperative to its members in case of emergency help enhance efficiency by increasing the effort of the members of the cooperative. The most important variables in the second factor are low participation in the board, frequent production planning, union membership, stable wages, and sanctions. These variables guarantee continuity of operations and a stable remuneration for labour input. The factor internal governance embraces stability of the board, daily wages, collective employment guarantee, and supervision. These aspects assure the members of stable benefits from collective actions. Information on the capital market is required, as fertilizers are often financed with credit.

Coinversion is an alternative to financing by credit. This option is becoming more important, as government credit facilities are now strongly restricted. The positive loading for the marketing variables indicates that more commercially oriented cooperatives are more familiar with the technical criteria for fertilizer application. A strict division of labour hinders compliance with these criteria. Services like collective marketing and frequent wage payment secure the associates' income. The incorporation of the union membership variable in the service factor is not surprising, as peasant organizations often render services to their members.

Efficiency in the use of *pesticides* is influenced by five factors: (i) clear rules for labour allocation and profit distribution; (ii) market information; (iii) cooperative services; (iv) planning procedures; (v) internal governance; The first three factors correspond with effects found for fertilizer efficiency. Union membership may serve as a substitute for information on the credit market, as peasant organizations are in charge of local NGO-financed programmes for credit distribution. A high incidence of family relationships apparently decreases the number of contacts and lowers the level of information. Planning procedures appear to be important to adapt to changing conditions regarding pests and diseases. Frequent planning of production and marketing and the permanent availability of labour on the collective plot guarantee timely control measures. Strict division of labour has the opposite effect. A stable board, daily wages and supervision provide certainty to the associates. Formal democracy – a high participation in the board – seems to cause loss of knowledge and experience. It is, however, important that the individual members can participate frequently in the cooperative's meetings and production planning. Union membership apparently provokes a decrease in self determination. The dictate of the members causes a large part of the profits to be distributed to the individuals. This clearly enhances efficiency in the short run, but in the long run profit prospects may be affected.

Efficiency in *tractor use* is dependent on three sets of variables: (i) clear rules for wage payment and profit distribution; (ii) internal assurance; (iii) cooperative services; and (iv) planning procedures. These factors were also found to favour the efficiency of other inputs and have therefore already been discussed.

Individual production

The allocative efficiency of labour is favoured by six factors: (i) services from the cooperative to its members; (ii) market information; (iii) payment rules; (iv) assurance mechanisms; (v) internal governance; and (vi) coordination. Post harvest services are not appreciated, as individual production is in general only sold in small units at critical periods, and concentrated labour efforts to reach deadlines for joint marketing affect flexibility of labour allocation. More effective land use and knowledge about co-inversion improve the labour intensity of production. The distribution of collective profits secures income flows for the individual peasants. The payment of membership subscriptions decreases the individual's funds. Assurance mechanisms also appear to be important for labour allocation. A relatively large amount of off-farm labour means a secure source of income outside the own plot. Long residence in the area is an indicator of established property rights. A high participation in the board decreases the time available for working on the own plot. Nevertheless, actual participation in the planning of collective production stimulates efficiency because it promotes the exchange of knowledge and experience. Coordination between a large number of associates who are members of many other organizations is also beneficial. However, labour exchange between associates may cause operations to be delayed during peak periods.

The use of *seed* in individual production could be improved by high stability of membership and is negatively influenced by post harvest cooperation and high participation rates in the board.

The variables that influence the use of *fertilizers* can be summarized in two factors: (i) internal assurance; and (ii) services and cooperation. The internal assurance factor consists of variables that guarantee a stable flow of income from collective production to the individual peasants: control of finance, stability of wages, emergency help, and sanctions. This assurance increases the willingness of the members to invest in collective production. As a result, they commit fewer resources to the individual plot. Labour exchange among members also has an adverse effect as it affects the flexibility of allocational decisions of farm households. Cooperation between associates and the provision of productive services (input purchase) favour fertilizer efficiency. Coordination between members is improved by frequent meetings. The admission of new members may (temporarily) disrupt cooperation.

The efficiency of the use of *pesticides* is influenced by three factors: (i) services from the cooperative to its members; (ii) coordination between associates; and (iii) market information. As expected, productive services favour pesticide use, but coordination between members (soil preparation, input purchase) decreases efficiency. Flexibility is extremely important for the control of pests and diseases and the moderate rigidity caused by cooperation outweighs the advantages. Besides, knowledge of land and capital markets enhances pesticide efficiency. Peasants who have been living in the region for a long period and who frequently work on the collective plot know more about the particularities of local pest control. As pest control mostly takes place outside the peak periods, labour exchange does not cause delays. It may even promote timely operations and exchange of information.

The variables that influence the use of *herbicides* are related to the internal organization of the APC and the flexibility of individual production. A large fraction of the land under collective production and internal assurance (frequent wage payment and a stable board) tie the individual's labour to the collective plot. Labour exchange between members also decreases flexibility.

The efficiency of *traction* use is decreased by the monitoring of labour on the collective plot, because strict supervision of labour on that plot can prevent timely soil preparation on the individually owned plot. Apparently, peasants with more individual assets have more scope for evading collective obligations. A second factor that influences the efficiency of traction is cooperative services. Union membership increases efficiency, as peasant organizations often rent equipment to their members. Collective transport decreases efficiency.

4   CONCLUSION

Management of common property resources has long been considered to be a device for inefficient allocation and resource depletion, as problems of free riding and of monitoring tend to promote non-cooperative behaviour. Within the property rights paradigm, external enforcement is required to overcome these problems, and thus privatization is promoted as a device to improve resource allocation according to opportunity costs. This approach is considered insufficient, as cooperation has emerged precisely as a substitute for missing markets. APC parcellation may 'resolve' boundary problems and behavioural uncertainty, but it tends to enhance market dependency and affects prospects for intertemporal and interspatial reciprocal exchange, and thus resource depletion is likely to increase.

Contract-choice theory offers new perspectives for the appraisal of feasible internal incentives and operational rules that can enhance efficient resource allocation and sustainable land use within a cooperative context. Using production function estimates and factor analysis we identified five different aspects of strategic interdependence among the cooperative enterprise and its members. Assurance and monitoring mechanisms based on detailed criteria for labour remuneration proved to be highly relevant for efficient resource use, while stable systems of internal governance and supply of APC services to members contributed equally to reducing strategic default. Moreover, access to market information, as well as direct coordination among associates is favoured by APC membership, even without full parcellation.

Problems of bounded rationality and asymmetries in decision making on resource allocation within APCs can be addressed through two different policy devices: (i) reform of internal organizational *rules* or (ii) establishment of *pricing mechanisms*. Honduran agrarian policies nowadays reflect a strong reliance on market forces, thus neglecting the importance of the internal assurance mechanisms based on reciprocity and interlinked exchange. Major doubts exist about the expected productivity-enhancing results of these policies, as market failures prevail. It may be argued that technical and allocative efficiency could be improved if major attention is given to relevant managerial variables of contract choice. Therefore, price instruments and market reform offer insufficient incentives, and institutional reform to reinforce allocational procedures for resource management should receive a higher priority.

Notes

1. A stable Nash equilibrium is defined as a situation where no agent has any incentive to deviate and select another strategy, because changing the actual strategy – given the strategic choice of other agents – offers no options for increasing pay-off. Individual contributions take everyone else's contributions as given.
2. Bartlett's test of sphericity indicates that the allocative efficiency scores for the different inputs are not completely independent. This implies that MANOVA should be applied to draw conclusions on the existence of overall differences in allocative efficiency. However, this is not possible, due to the small number of cases for collective production units. Fortunately, there is no danger of overlooking differences, because most univariate tests give statistically significant scores.

Bibliography

Berg, M. van den (1995) *Economic Efficiency of Agricultural Production Cooperatives in Comayagua*, Honduras, MSc. Thesis, Wageningen Agricultural University.

Bromley, D.W. (1992) 'Land and Water Problems: An Institutional Perspective', *American Journal of Agricultural Economics* (64) pp. 834–44.

Bromley, D.W. (ed.) (1992) *Making the Commons Work: Theory, Practice and Policy* (San Francisco: ICS Press).

Carter, M.R., J. Melmed-Sanjak and K.A. Luz (1993) 'Can Production Cooperatives resolve the conundrum of exclusionary growth? An econometric evaluation of Land Reform Cooperatives in Honduras and Nicaragua', in C. Csaki and Y. Kislev (eds) *Agricultural Cooperatives in Transition* (Boulder: Westview Press).

Douma, S. and H. Schreuder (1991) *Economic Approaches to Organizations* (Hempstead: Prentice-Hall).

Feeny, D., F. Berkes, B.J. McCay and J.M. Acheson (1990) 'The Tragedy of the Commons: Twenty-Two Years Later', *Human Ecology* (18), 1 pp. 1–19.

Fekete, F., E.O. Heady and B.R. Holdren (1976) *Economics of Cooperative Farming: Objectives and Optimum for Hungary* (Leiden/Boedapest: Sijthoff/ Akadémia Kiadó).

Hair, J.F., R.E. Anderson, R.L. Tatham and W.C. Black (1992) *Multivariate Data Analysis* (New York: Macmillan Publishing Company).

Hardin, G. (1968) 'The Tragedy of the Commons', *Science* (162), pp. 1243–8.

Hayami, Y. and K. Otsuka (1993) *The Economics of Contract Choice: An Agrarian Perspective* (Oxford: Clarendon Press).

Larson, B.A. and D.W. Bromley (1990) 'Property Rights, Externalities, and Resource Degradation', *Journal of Development Economics* (33), pp. 235–62.

Oakerson, R.J. (1992) 'Analyzing the Commons: A Framework', in D.W. Bromley (ed.) *Making the Commons Work: Theory, Practice and Policy* (San Francisco: ICS Press).

Olsen, M. (1971) *The Logic of Collective Action* (Cambridge: Harvard University Press).

Ostrom, E. and R. Gardner (1993) 'Coping with Asymmetries in the Commons: Self-Governing Irrigation Systems can work', *Journal of Economic Perspectives* (7), 4 pp. 93–112.

Platteau, J.P. (1992) *Land Reform and Structural Adjustment in sub-Saharan Africa: controversies and guidelines*, Rome: FAO Economic and Social Development Paper No. 107.

Putterman, L. (1981) 'On Optimality in Collective Institutional Choice', *Journal of Comparative Economics* (5), pp. 392–402.

Ruben, R., C.J. Wattel, F. Funez and M. Ponce (1992) *Perspectivas para la Consolidación Empresarial del Sector Reformado en Honduras*, San José: CDR-VU Field Report.

Ruben, R. and P. Roebeling (1995) *Options for the Consolidation of Land Reform Cooperatives: a Contract-Choice Approach*, Paper presented to Congres 'Agrarian Questions', Wageningen Agricultural University, May 1995.

Runge, C.F. (1984) 'Strategic Interdependence in Models of Property Rights', *American Journal of Agricultural Economics* (66), pp. 807–13.

Schlager, E. and E. Ostrom (1992) 'Property-Rights Regimes and Natural Resources: A Conceptual Analysis', *Land Economics* (68) 3, pp. 249–62.

Wade, R. (1987) 'The Management of Common Property Resources: Collective Action as an Alternative to Privatization or State Regulation', *Cambridge Journal of Economics* (11), pp. 95–106.

11 Land Titling and Prospects for Land Conservation: Lessons from a Case-Study in Honduras
Daniel Wachter

1 INTRODUCTION

In recent years, the development debate has increasingly focused on institutions and the legal environment in developing countries (World Bank, 1991). The role of secure property rights as a basis for entrepreneurial activities has received particular emphasis. De Soto (1989) became widely known for his research in Peru, in which he argued that an inefficient bureaucracy complicates the obtaining and the exchange of legal property rights, forcing the poor majority of the population into the informal sector. The resulting insecurity of property rights is thought to reduce incentives for investment because, in the informal sector, economic agents cannot be certain they will reap the fruits of their own efforts. Whereas de Soto concentrated on property rights issues in an urban environment, other authors like Johnson (1972) or Feder *et al.* (1988) investigated property rights in rural areas and in relation to stagnating agriculture. Recently, the growing environmental crisis has also enhanced interest in property rights.

In many developing countries, land rights are unclear, unspecified, or disputed, and legal titles are often lacking (Leonard, 1987; Southgate, 1988). While in the African context the property rights debate refers mostly to traditional systems of communal land management, in Latin America the problem of tenure insecurity, in general, is related to inequality of landownership, which weakens the economic and legal position of poorer farmers and pushes them onto small, marginal plots that are more vulnerable to land degradation. In Honduras, the problem also has to do with the high proportion of national/*ejidal* land that has been invaded and is now used by individual farmers. Missing, attenuated, or insecure property rights are often thought to be an important cause of resource degradation because they reduce the incentive to conserve resources. Many have therefore advocated the establishment or strengthening of exclusive property rights. Land titling and the provision of legal titles to farmers are thought to play an important part in reducing problems related to insecure tenure and, consequently, in encouraging land conservation (Feder and Feeny, 1991; Lemel, 1988).

As far as empirical data are concerned, there is plenty of information on the effects of land rights registration on agricultural investment in general (Dickerman, 1987; Feder *et al.*, 1988). However, the effects of land rights registration and land titling on conservation have hardly been investigated. The author of this paper, during a research stay with the Environmental Policy and Research Division of the World Bank in Washington, DC, in 1991–92 had the opportunity to visit a USAID (US Agency for International Development) land titling project in Honduras and to analyse its effects on soil conservation (Wachter, 1992b).

2　THE RATIONALE OF LAND TITLING FOR LAND CONSERVATION

Traditionally, economists have considered investments in conservation to be analogous to investments in enhancing the productivity of land. Investments in conservation might serve to prevent reductions in future income streams (Collins and Headley, 1983), to increase future income streams (Feder *et al.*, 1988), or to increase the value of land as a capital asset (King and Sinden, 1988; Palmquist and Danielson, 1989). When property rights are missing or insecure, however, economic agents cannot be sure they will receive the benefits of their efforts and, therefore, have little incentive to invest. To the extent that investment does occur, the planning horizon and duration of investments tend to be short term (Johnson, 1972). Exclusive property rights would restore farmers' incentives to invest in their land. For this to occur, it is quite important that property rights include both use and transfer rights, as well as rights to obtain income from the asset.

Land titling can be defined either as the act of assigning rights (whether formal or informal) or as the provision of legal recognition to existing rights or interests. Here, the term is confined to the latter sense (Dale and McLaughlin, 1988). The rationale for land titling rests on three objectives. First, and probably most fundamental, is the objective of increasing the security of tenure proper. Registered or titled land rights are thought to be more secure than unregistered rights because they are guaranteed by the state in the case of conflict. Feder *et al.* (1988, p. 28), for example, explicitly equate secure tenure with legal title. 'Security of ownership is defined ... as the possession of legal rights of ownership, certified by an appropriate state-issued document.' Promoters of land titling argue that the link between secure tenure and legal title makes land titles a necessary prerequisite for investments in farm productivity and land conservation.

The second objective of land titling is to increase the demand for and supply of credit. Many investments in farm productivity and land conservation require capital inputs. Lack of clear legal title is thought to prevent farmers from using their land as collateral for credit. This is particularly important in the formal credit market, where lenders rarely have personal or detailed information about prospective borrowers. Although titles are typically less significant in the informal credit market, informal credit tends to be much more expensive than formal credit and to be confined to relatively small, short-term loans (Aleem, 1990). The more that farmers' incentives to invest increase, the greater the farmers' demand for credit (Roth and Barrows, 1988).

The third objective of land titling is to foster land markets. The classic economic argument for making land a tradeable item is that it promotes transfers of land to more productive – as well as soil conserving – farmers. It is widely believed that legal land titles are an essential prerequisite for the working of land markets (Stringer, 1989; Feder and Feeny, 1991). Legal land titles facilitate land transactions by reducing information costs and uncertainty about land rights. Without legal title, potential buyers of a piece of land cannot be certain that they are buying land from its real owners, or they have to incur high costs in order to obtain full and accurate information. Poorly functioning land markets tend to lower land values, other things being equal, because effective demand is limited. This can affect farm productivity and land conservation in several ways (Johnson, 1972). First, incentives for conservation are reduced because owners cannot realize the benefits of investments if they sell the land. Second, low land values reduce the value of land as collateral, since the lender cannot easily sell the land to recover the lost credit. Therefore, when land markets function poorly credit tends to be more expensive and investment is reduced.

3   THE USAID LAND TITLING PROJECT IN HONDURAS

In Honduras population growth pushed an ever increasing part of the agricultural population into former forest areas owned by the state and the local municipalities where private property is prohibited, on principle. For decades the state tolerated informal, de facto, private ownership. Then, in 1982 in collaboration with USAID, the Government of Honduras started to make some of these farmers legal by means of a land titling project (Proyecto de Titulación de Tierra, PTT). The land titling project aimed at legalizing the informal property rights of 40 000 farmers in seven departments of the country, increasing these farmers' security of tenure, facilitating the land market and the use of land as collateral for credit. An implicit objective was to expand coffee as a hillside crop, because this perennial crop is thought to be beneficial for hillside management. PTT is one of the best documented land titling projects because, from its outset, parallel investigations by researchers from the Land Tenure Center of the University of Wisconsin-Madison were planned (Nesman and Seligson, 1988; Stanfield *et al.*, 1990).

4   THE CASE STUDY IN HONDURAS

The evaluation of the Land Tenure Center researchers was focused on the general investment behaviour, particularly of the numerous coffee growers, but neglected questions of land degradation and soil conservation. Therefore, this author aimed at analysing land degradation and soil conservation in his empirical case study in the Santa Barbara department in the northwest of the country, where PTT started in 1982 and where results should already be observable. The objectives of the case study were, first, to reconsider the results of previous evaluations of land titles and general investment behaviour, and second, to study differences between titled and untitled farmers concerning land degradation and soil conservation.

Working closely with the USAID mission in Honduras and the Honduran National Agrarian Institute (INA), 65 farmers were selected and interviewed. In semi-standardized interviews, information was gathered on the farm (e.g. whether coffee growing or not), on the perceived pros and cons of land titles, on erosion problems and soil conservation measures, and on the use of chemical inputs.

Like most departments of Honduras, Santa Barbara is very mountainous, and many steep deforested areas are cultivated (while the flat areas are used for extensive livestock farming). However, there is no dramatic erosion. Of the 65 interviewed farmers, 24 were confronted with serious erosion problems. It would have been possible to locate the case study in other departments with more erosion problems, but there, the PTT had not been under way for long enough for differences between titled and untitled farmers to have emerged.

In the PTT, land titling was encouraged but not compulsory. A farmer wishing to obtain an INA land title had to apply for it. Three neighbors had to attest to the merit of his claim, to ensure that outsiders could not grab the land. 44 of the 65 peasants interviewed (68 per cent) had an INA land title and 21 (32 per cent) did not. The average holding of title-holders was 14.7 ha (ranging from 0.7 ha to 182 ha), 16.7 ha for titleless farmers (ranging from 0.9 ha to 90 ha). 70 per cent of the title-holders were coffee growers (because it is a perennial crop, coffee is beneficial for soil conservation), but only 43 per cent of them had more than half of their land under coffee. 81 per cent of the titless farmers grew coffee, only 33 per cent of them on more than half of their land.

So there were two groups of farmers available with more or less identical attributes except for the title status. It must be stressed that most of the farmers were illiterate. The information they gave was usually not very precise, and therefore it was not suitable for sophisticated statistical analyses.

5 GENERAL EFFECTS OF LAND TITLES

A general finding from the interviews was that title ownership had only a limited effect on the farmers' behaviour and that the causal direction was unclear, that is it was not clear whether a land title increased investment or whether a title to the land was used to secure a previously planned or executed investment. This case study thus confirmed an evaluation of the PTT in 1988:

'The findings are that land titles are likely to be necessary but not sufficient ... Whereas previous studies have suggested that agricultural development programs are constrained by insecurity of tenancy, this study shows that a tenure security program not combined with systematic efforts to deliver key inputs, especially credit ... has little impact' (Nesmann and Seligson, 1988, p. iii).

Before the PTT, tenure was not really insecure. There were hardly any land conflicts, farmers respected each other's land ownership. Informal property rights and their local recognition proved to be more important than legal title. Additionally, contrary to expectations, land rights were not always completely informal. The farmers often possessed documents issued by local municipalities, or private sale contracts (*escrituras privadas*). Thus the national land titling project replaced these local arrangements with a new, national land titling and registration system.

Therefore, the principal effect of the PTT could not be an increase in tenure security proper, but facilitating the land market and the access to credit. In reality, an INA land title changed the farmers' situation only gradually in this respect too, because:

- even without INA land title, coffee growers had access to credit from IHCAFE, the state owned Honduran Coffee Institute (3 of the 21 titleless farmers),
- contrary to expectations, banks accepted the local *escrituras privadas* as ownership documents (a further 3 of the 21 titleless farmers),
- in some cases the banks accepted other securities like a house or the harvest (yet another 3 of the 21 titleless farmers).

So as many as 43 per cent of this sample of farmers without INA land title had access to formal credit, that is without having to have recourse to informal lenders offering expensive credit. On the other hand, *minifundistas* with or without INA land title had difficulty in getting credit, in any case, because the banks did not lend sums less than 5000 Lempiras (US\$ 1000) in 1992.

In this respect, an information problem on the farmers' part was revealed. At the outset, the land title showed only a symbolic land value, so as to make the land titles more affordable for the farmers (land titles were not given away for free, but had to be bought). Thereafter, land improvements such as planted coffee trees could be registered, thereby increasing the land value and permitting higher credit; 10 of the 44 title-holding farmers were not informed about the necessity to have land improvements registered. Furthermore, some farmers mentioned that they had not applied for credit because they could not afford the interest.

In spite of these problems, the majority of the farmers regarded INA land titles very favourably. An INA land title seems to have psychologically increased the feeling of security, even though no major changes of behaviour were found. 13 of the 21 titless farmers (62 per cent) said they were interested in acquiring a land title. Asked why they had not applied in the first place, most indicated that they had done so, but had not received the title. There seem to be bureaucratic reasons for this, such as applications lost in INA headquarters in Tegucigalpa or title lost during delivery. However, the titless farmers' interest in land title can also be explained in terms of the banks' changed policy on securities. The banks seem to be putting more and more emphasis on INA land titles in their lending operations, that is the need or benefit of having an INA land title has increased.

6 EFFECTS ON SOIL CONSERVATION AND USE OF CHEMICAL INPUTS

As indicated earlier, 24 of the farmers interviewed were confronted with serious erosion problems; 16 of them were title-holders. They were asked whether they used one or more of the following conservation techniques:

(i) physical conservation measures, e.g.:
   - construction or maintenance of terraces,
   - stone walls,
   - tree alleys,
   - other biological conservation techniques,
(ii) specific land use techniques (e.g. contour farming).

Seven of the eight titleless farmers with erosion problems (87 per cent) used at least one of the physical conservation, techniques mentioned; most preferred simple and cheap control measures like stone walls and planted erosion barriers. Two farmers constructed or maintained terraces. Of the title-holders with erosion problems, 12 of 16 (75 per cent) used one or more conservation techniques, again cheap and simple ones. The question about specific land use techniques provided a wide range of imprecise answers that were difficult to categorize. Only few farmers – in both groups – explicitly used soil conserving techniques.

As to the use of chemical inputs (fungicides, pesticides, herbicides, fertilizers) again the imprecise answers have to be mentioned. Reduced to a simple yes/no question, that is whether or not they used chemical inputs at all, hardly any differences between title-holding and titleless farmers were found; 20 of 44 title-holders (45 per cent), 9 of 21 titleless farmers (43 per cent) used chemical inputs.

Summing up, titleless farmers, contrary to expectations, tend to use physical erosion control measures slightly more than title-holders, and the use of chemical inputs in both groups is more or less identical. However, considering the rather small sample, a more appropriate conclusion would be that there are no significant differences in general investment behaviour or in soil conservation activities between title-holding and titleless farmers.

7 CONCLUSIONS

(i) Land titling changes the property rights situation only gradually

Property rights situations are complex. The often depicted contrast between clear, specified property rights on the one hand and non existent property rights on the other is too simplistic. Lack of an official land ownership document does not necessarily imply tenure insecurity. Informal rights may be fairly secure. In this case study, farmers respected each other's land rights before INA land titling. In addition, farmers had semi-official documents issued by the local municipalities. Furthermore, there is no guarantee that in the case of land conflict a land title really increases the security of tenure, if the state lacks the powers of sanction. So in this case study, land titling changed the property rights situation only gradually, replacing local and semi-official arrangements with a national system.

(ii) Land titles are not sufficient for sustainable use of soils

The property rights literature suggests that clear property rights/land titles are the key factor for sustainable use of soils. This study showed, however, that land titles are welcomed and appreciated by farmers, but that other factors such as credit and agricultural extension are at least as important. Isolated land titling projects, not accompanied by other rural development activities are unlikely to achieve their objectives.

(iii) Land conservation measures are assessed in terms of costs and benefits; cheap and simple techniques are preferred

The soil conservation techniques used by the farmers were simple and cheap (stone walls, planted barriers), in general. This confirms that conservation measures are assessed in terms of costs and benefits. This means that farmers will not use every technique, even if property rights are formal, clear and specified, if the costs exceed the benefits. Therefore, extension and research on soil conservation should pay attention to these economic aspects and promote cost-effective conservation techniques.

(iv) An inefficient bureaucracy reduces the potential benefits of land titling programmes

In this case study it was shown that the executing agency, INA, did not fulfill its tasks in a satisfactory manner (such as insufficient information of farmers, slow treatment of title applications). INA thus reduced the potential benefits of PTT. Therefore future land titling programmes will have to pay more attention to institutional development of the executing agencies.

Bibliography

Aleem, I. (1990) 'Imperfect Information, Screening and the Costs of Informal Lending – A Study of a Rural Credit Market in Pakistan', *World Bank Economic Review* 4(3), pp. 329–49.

Collins, R.A. and J.C. Headley (1983) 'Optimal Investment to Reduce the Decay Rate of an Income Stream: The Case of Soil Conservation' *Journal of Environmental Economics and Management*, 10, pp. 60–71.

Dale, P.F. and J.D. McLaughlin (1988) *Land Information Management* (Oxford: Clarendon Press).

De Soto, H. (1989) *The Other Path* (New York: Harper & Row).

Dickerman, C. (1987) (Hrsg.) 'Security of Tenure and Land Registration in Africa: Literature Review and Synthesis', LTC Paper no. 137 (Madison: Land Tenure Center, University of Wisconsin).

Eckholm, E.P. (1979) 'The Dispossessed of the Earth: Land Reform and Sustainable Development', World watch Paper no. 30 (Washington, DC: World watch Institute).

Feder, G. and D. Feeny (1991) 'Land Tenure and Property Rights: Theory and Implications for Development Policy', *World Bank Economic Review*, 5(1), pp. 135–53.

Feder, G., T. Onchan, Y. Chalamwong and C. Hongladarom (1988) *Land Policies and Farm Productivity in Thailand* (Baltimore: Johns Hopkins University Press).

Johnson, O.E.G. (19/2) 'Economic Analysis, the Legal Framework and Land Tenure Systems', *Journal of Law and Economics* 15(April), pp. 259–76.

King, D.A. and J.A. Sinden (1988) 'Influence of Soil Conservation on Farm Land Values', *Land Economics*, 64, pp. 242–55.

Lemel, H. (1988) 'Land Titling: Conceptual, Empirical and Policy Issues', *Land Use Policy*, July, pp. 273–90.

Leonard, H.J. (1987) *Natural Resources and Economic Development in Central America* (New Brunswick, NJ: Transaction Books).

Nesman, E. and M. Seligson (1988) 'Land Titling in Honduras: an Impact Study in the Santa Barbara Region', Report to the United States Agency for International Development, Washington, DC.

Palmquist, R.B. and L.E. Danielson (1989) 'Erosion, Drainage and Land Values', *American Journal of Agricultural Economics*, 71, pp. 55–62.

Quiggin, J. (1987) 'Land Degradation: Behavioural Causes', in A. Chisholm and R. Dumsday (eds) *Land Degradation – Problems and Policies* (Cambridge: Cambridge University Press).

Roth M. and R. Barrows (1988) *A Theoretical Model of Land Ownership Security and Titling Impacts on Resource Allocation and Capital Investment* (Madison: Land Tenure Center, University of Wisconsin).

Southgate, D. (1988) 'The Economics of Land Degradation in the Third World', Environment Department, Working Paper, no. 2, Washington, DC; World Bank.

Stanfield, D., E. Nesman, M. Seligson and A. Coles (1990) 'The Honduras Land Titling and Registration Experience' (Madison: Land Tenure Center, University of Wisconsin).

Stringer, R. (1989) 'Farmland Transfers and the Role of Land Banks in Latin America', LTC Paper no. 131 (Madison: Land Tenure Center, University of Wisconsin).

Wachter, D. (1992a) 'Farmland Degradation in Developing Countries: The Role of Property Rights and an Assessment of Land Titling as a Policy Intervention', LTC Paper no. 145 (Madison: Land Tenure Center, University of Wisconsin).

Wachter, D. (1992b) 'Die Bedeutung des Landtitelbesitzes für eine nachhaltige landwirtschaftliche Bodennutzung – eine empirische Fallstudie in Honduras', *Geographische Zeitschrift*, 3, pp. 174–83.

Wachter, D. (1996) 'Land Tenure and Sustainable Management of Agricultural Soils', Development and Environment Reports, no. 15 (Berne: Group for Environment and Development, Institute of Geography, University of Berne).

Wachter, D. and J. English (1992) 'The World Bank's Experience with Rural Land Titling', Environment Department, Divisional Working Paper, no. 35 (Washington, DC: World Bank).

World Bank (1991) *World Development Report 1991 – The Challenge of Development* (Washington, DC: Oxford University Press).

World Bank (1992) *World Development Report 1992 – Development and the Environment* (Washington, DC: Oxford University Press).

12 Non-Conventional Rural Finance and the Crisis of Economic Institutions in Nicaragua

Johan Bastiaensen

1 INTRODUCTION[1]

Nicaragua could become one of the few Latin American countries with an equitable rural development path. Since 1983, different phases of land reform have reduced the property share of large estates from about 36 per cent to just over 10 per cent and have transferred it to small-scale producers (INRA-OEA, 1991). Further, economic analysis has shown that smaller scale peasant production systems provide social value added and net generation of foreign exchange at least equal to that of most larger scale entrepreneurial farms (Bastiaensen, 1994; Maldidier, et al., 1995). Under today's conditions of structural adjustment one could thus expect peasant systems to thrive. However, a quick glance at reality shows that such optimism is not warranted. One of the reasons for the disheartening state of affairs is the crisis of the economic institutions. War and harsh political struggle as well as the transformation of the agrarian structure have disrupted the country's socio-economic fabric. At present, one can but observe the incompatibility between the relatively equal distribution of land and the shattered remnants of the previous economic institutions. As yet, no strong new and more appropriate institutional structures have emerged. However, as M. Lipton (1994, p. 642) affirms, land reform requires alternative institutions to 'replace (or dispense with) the endogenously required economic functions of the old structures ("rural tyrants" as sources of loans, security and technologically innovative risk-taking or "informal magistracy")'. Institutional deficiencies therefore constitute an important bottleneck for economic recovery in the rural sector in the post-agrarian reform era.

191

2 ELEMENTS OF THE INSTITUTIONAL ENVIRONMENT IN THE COUNTRYSIDE

A macro-view of the overall socio-economic articulation and the present institutional crisis

D. Kaimowitz (1989, p. 50) rightly argued that the Somozista economy constituted a relatively well articulated system. In the rural areas a landed oligarchy maintained control over a substantial part of the rural economy. Oligarchical control over land and other resources was crucial to guarantee a sufficient supply of seasonal labour for the coffee, sugar and cotton crops. This labour was tied to the estates through simultaneous credit, share-cropping and commercial contracts as well as personal obligations originating from specific favours and occasional assistance within a framework of paternalistic relations. Much of rural finance, provisioning and commercialization also passed through the large landowners. Vertical patron–client relationships were the governance structures of economic transactions. Typical of the agrarian structure of Nicaragua and diverging somewhat from the typical Latin American dual structure was the presence of a substantial number of peasants with small and medium-sized farms, particularly in the agricultural frontier regions (Barraclough and Marchetti, 1985, p. 191; Kaimowitz, 1987, pp. 24–5). In the latter areas, where the oligarchy was often absent, its functions were assumed by successful producers of peasant origin, *finqueros*. Social articulation followed similar patron-client patterns, which were ideologically even more strongly rooted, because of the closer cultural proximity (CIERA, 1989, pp. 296–303).

During the Sandinista revolution, the rural socio-economic institutions disarticulated. The *somozista* segment of the landed oligarchy was replaced by state enterprises and the economic control of the non-*somozista* bourgeoisie undermined. For several years, 'free market' provisioning and marketing channels were largely prohibited and replaced by a deficiently functioning system of state enterprises and administered prices. In the process, the previous interlinked credit, labour and land market relations were largely dissolved and not replaced by a viable alternative. Disruption was great, especially in the remote peasant areas, and can be considered both the cause and consequence of the emergence and consolidation of the rural *contra guerilla*.[2]

Today, the Nicaraguan rural economy is struggling to meet the challenge to find a viable institutional rearticulation. To date, liberalization of trade has put an end to ineffective state interference but has merely resulted in a severe contraction of the outreach of financial and commodity markets altogether. Additional land reform was organized under the demobilization process, so that the landed property structure has been irreversibly equalized. Thus a simple return to the previous oligarchical governance structures can be ruled out. However, despite ten years of revolution and massive NGO presence, not many new democratic structures are emerging. The revival of old vertical structures and the creation of new ones, in which the remaining landowners and the more accommodated *finqueros* reassume their previous roles, seems more probable than a transition towards more horizontal economic structures. Still, many crucial issues of the institutional rearticalation such as property rights, the nature of financial and commercial configurations and local governance structures remain uncertain. Today, the Nicaraguan countryside mainly shows signs of a profound and unresolved institutional crisis.

The micro-level institutional environment in the countryside

In most peasant territories in Nicaragua, the extended family remains the basic social relationship. The primary family network is usually supplemented by a more or less developed personal network beyond the lineage. Exchange and solidarity within these networks will not be as strong as within the lineage, but nevertheless constitutes an indispensable day-to-day reference for economic and social interaction in the local territory. Within these networks local leaders play an important role. Their legitimacy and power depends crucially on their capacity to intermediate with the outside world (Marchetti, 1994, p. 6). For their access to government and outside markets, clients often depend to a large degree on their patron. Important outside connections can be the state, certain private enterprises, political parties, interest organizations, churches and NGOs. Until the Sandinista revolution, dominant patron-client networks were often identified as 'conservative' or 'liberal-*somozista*' according to the two fractions of the dominant oligarchy. Within the networks, relations are traditionally vertical and clientlistic. Leaders behave like *caudillos*, that is they exercise local power in an authoritarian way. They may sometimes be responsive to the needs of their clients, but are hardly ever accountable to their 'constituency'. Fraud for the personal benefit of local leaders is therefore very common, as is shirking on the part of the dependent clients.

In most cases, the Sandinista revolution has not changed the fundamental nature of these local relations. New parallel local power structures have developed, or old ones have been transformed, and have become linked to party-state organizations. In this way, the prospects for political emancipation for new and (sometimes) old locally dominant groups have improved. Many 'conservative' groups have switched to 'sandinista'. Other, often 'liberal-*somozista*' groups' has drastically decreased so creating the ground for stronger bonds with oppositional structures like churches, business organizations or the *contra* military. A feature of the Sandinista political system with its paternalistic socialist state, was that it enhanced 'rent-seeking', that is the search for benefits from redistribution through the political process. This strengthened local political clientelism rather than enhancing autonomous horizontal organization for local problem solving. Internal local power structures also remained extremely vertical and authoritarian, with local *caudillos* strategically linked to the outside structures of the revolution. Marchetti indicates that the practices of social dependency and clientelism were so legitimate among the 'bases' that one could consider them as 'the traditional peasant democracy' because of the broad and spontaneous consensus that they evoked (Marchetti, 1994, p. 6). This terminology also draws attention to the inherently ambiguous nature of the local social relations that express both dependency/subordination as well as identification/community. An important further observation is that this underlying social reality is quite resistant to national (externally induced) political change. 'The traditional consensus reproduces itself under the umbrella of whatever political or ideological rethoric.' (Marchetti, 1994, p. 6).

Today, peasant communities in Nicaragua are generally composed of several competing patron-client networks, unless they are situated in the more homogenous agricultural frontier villages where often there is just one dominant network present. Usually, the different social sectors scarcely interact with each other. Quite often relationships are conflictive. The Sandinista revolution and the war that was imposed on the country greatly exacerbated local polarization and conflict.

In this context, however, an important hypothesis is that these political divisions in the local territory above all express historical differentiation of local family and patron-client networks and less so any real political or religious differences on national issues. The latter seem more a result of adaptive strategic alliances between relatively stable local structures and national organizations that are contingent on the evolution of the political scene. In the context of the recent disillusionment with national politics, this also allows for something surprising and positive. Despite recent political polarization and harsh (military) confrontation, local détente often proves to be easier than expected. Many people in rural communities, especially in the war zones, seem to have a profound feeling of being abused for interests other than their own (Bendaña, 1991, p. 108). This feeling might create motives for reducing exhausting political confrontation and create the conditions for local pacification as well as the necessary space for community-based solutions to local problems. The tendency is further enhanced by the severely reduced subsidies that are offered to the rural areas by outsiders and the state. This imposes the need to look for local rather than external solutions.

The institutional change required for equitable rural development

Rural development in Nicaragua faces two challenges. The first is the need to at least pacify local relations and, if possible, to enhance local co-operation between the different social sectors. As indicated above, present external conditions might be supportive of this. The second is the need to rearticulate the local governance structures within social networks so that economic transactions can develop in a more appropriate institutional framework.[3] The latter has to take account of the effect of the radical change in the landed property structure. With the resulting substantial increase in the number of rural producers, this increasingly invalidates vertical governance structures as a framework for economic transactions. Only more horizontal local cooperation between and within social networks can create and support the necessary virtuous circles of increasing mutual confidence, improved communication, collective learning and cooperation on the basis of which both the local productive capacity and the insertion into the wider economy can be improved. Diversified horizontal interaction can therefore be held most compatible with the uni-modal agrarian structure. Vertical mechanisms of force and dependency lack these kinds of virtuous circles and are much more conducive to stagnation and fraud (Putnam, 1994, p. 70).

In Nicaragua, a transition towards a rural 'civic community' of cooperation based on self-interest will be critical for the prospects of equitable development under the present-day agrarian structure.[4] Given the intrinsic vertical structuring of power in Nicaragua, a cultural transformation of the basic economic institutions in this direction is therefore required in order to enhance prospects for equitable rural development. The legacy of domination and verticalism is, however, present and it would be quite unrealistic to assume it did not exist. Therefore one of the main challenges for alternative rural development is to articulate operational strategies that enhance such a democratic transition. In the next section this question will be elaborated in the specific context of rural financial market development.

3   RURAL FINANCE AND INSTITUTION BUILDING FOR EQUITABLE RURAL DEVELOPMENT

The present-day rural financial configurations in Nicaragua

Nicaragua is characterized by a strong duality in financial configurations. Under the impact of recent stabilization policies, one can also observe a privatization and liberalization process as well as a severe contraction of the formal financial markets and credit itself. As a percentage of national agricultural product, rural credit supply fell from over 40 per cent in the period 1983–7 to 16 per cent in 1993.[5] As can be read from Table 12.1, the state development bank BANADES remained by far the single most important source of rural finance for both large and medium scale pro-ducers. Three quarters of total supply derives from this source and reaches one quarter of the total clients. Still, outreach compares fairly negatively with the previous decade when BANADES covered over 80 000 rural clients. The smaller rural clients in particular have been excluded. After 1993, BANADES outreach decreased further. In 1995, another 23 of the remaining 66 rural branches were scheduled to disappear (Doligez, 1994, p. 4). Contraction of BANADES finance has much to do with the poor repayment record of this financial institution. In the early

Table 12.1   Nicaragua: Institutional rural credit by source and destination in 1993 (millions of dollars)

Source and destination of credit	Amount	%	Clients	%	Average Amount
Large entrepreneurs	62.70	65.0	2100	3.8	29 857
BANADES	46.00	48.0	2020	3.7	22 772
Government IRD projects	9.00	9.0	n.a.	–	n.a.
Private banks	7.70	8.0	80	0.1	96 250
Small & Medium Producers	33.50	35.0	52 945	96.2	633
BANADES	25.00	26.0	10 815	19.6	2312
Government IRD projects	5.10	5.0	18 904	34.4	270
Sustainable NGO finance	2.60	3.0	16 020	29.1	162
Assistential NGO finance	0.80	1.0	7206	13.1	111
TOTAL RURAL CREDIT	96.20	100.0	55,045	100.0	–

Source: Nitlapán (1944b, p. 23).

1990s repayment rates oscillated around 70 per cent. The legacy of previous political remissions means that each economic or climatological problem almost immediately leads to unwillingness to pay and demands for flexibility. On the other hand, legal enforcement – and particularly the sale of collateral (land) – has hardly been possible up to now. Of course, capacity to pay has also been affected by the high real interest rates of over 20 per cent per annum.

Government integrated rural development projects are a second important source of institutional credit. These represent 14 per cent of total credit supply. IRD projects are particularly important for the smaller producers for whom they are the most important source of credit. Repayment of smaller client ranges between 35 per cent and 45 per cent. In this targeted project-credit, donor pressure for more efficiency is also gradually eroding the possibility of continuing to subside deficits.

Financial liberalization has led to new private banks also appearing on the scene. It was hoped that they would take over the space left by the contracting state banks. Private banks, however, delivered a poor 8 per cent of total credit supply, concentrated on a small number of very large clients with average loans close to the equivalent of 100 000 dollars. The repayment record is assumed to be slightly better than in BANADES, with repayment rates around 80 per cent.

The small rural producers, never able to access formal credit or suffering the retreat of BANADES, are increasingly served by NGO financial projects or programmes. An estimated 4 per cent of the total credit supply has already been derived from this source, serving some 25 000 small clients with micro-credits. The largest share comes from 'sustainable' NGO-sources that strive for profitable finance and therefore emphasize cost containment and repayment. In general, their recuperation rates would already be over 80 per cent. A minor share of NGO credit is targeted at the very poor and has a more paternalistic approach. Their repayment would be less than 50 per cent, implying a high degree of subsidies and enterpreneurial inviability.

It is clear that in the present constellation a substantial part of the rural area has been left with a very deficient access to finance. A recent survey of Nitlapán in the interior regions[6] found that only 18 per cent of producers had access to any source of credit (Nitlapán, 1995, p. 38). The need for rural financial configuration development is evident. Its main problem is institutional: how to intermediate financial resources to the majority of small and medium-sized rural producers in an entrepreneurially sustainable way.

An institutional perspective on sustainable rural finance

The challenge of viable non-exclusive rural finance consists in the creation of a stable organizational framework in which financial operations with small and medium-sized rural producers can take place-both in profitable conditions for the financial enterprise ('bank') and without excessive costs for the envisaged clients. This requires appropriate governance structures that are largely absent in the Nicaraguan countryside. Conventional banking procedures imply that positive banking profits would require prohibitively high interest rates to cover the high administrative costs of relative small loans as well as to ensure some repayment through expensive and often ineffective legal procedures. It should thus be no surprise that repayment problems in present-day rural finance are severe and that conventional rural finance is retreating.

The control of transaction costs is a prerequisite for any viable institutional alternative for rural finance. Four types of transaction costs can be distinguished. Two are *ex ante*: the costs of information and negotiation; two are *ex post*: the enforcement and adaptation costs. Since information and enforcement costs are most important, this paper will focus on these. The whole of transaction costs relates to the specific nature of the interface between the two transacting parties: the bank and the clients. Viable rural finance configurations thus have to define and sustain a cost-minimizing organizational framework that governs the relation between the financial enterprise and its clients. This bank–client interface should preferably also contribute to the strengthening of the broader local institutionality in order to increase the long term development opportunities for the bank's clients (see below).

Rural finance and information costs

A first component of transaction costs are information costs. The problem at hand is the selection of clients given the presence of information asymmetry between the bank and the client. For a conventional financial enterprise, it is very difficult and costly to determine the viability of projects as well as the overall solvency of small rural producers. It is hardly possible to evaluate the honesty and seriousness (or for that matter the drinking habits) of a potential client. These personal attributes are, however, crucial to predict future 'willingness to pay'. The selection of clients therefore requires a screening mechanism at the local level where this kind of information is more easily available.

Two broad non-exclusive alternatives can be envisaged. The first is to use the knowledge of the local leaders. They can be assumed to have excellent knowledge of 'their people'. In the Nitlapán system of non-conventional finance, a selection committee formed by local leaders does the selection job. This provides for a screening mechanism that articulates relatively easily to the prevailing vertical governance structures in the Nicaraguan countryside. If not complemented by other mechanisms (see 2.3.), it will however share the disadvantages of the vertical framework in terms of incentives for fraud (leaders) and shirking (clients), distorted information flows and unequal imposition of norms and sanctions.

The second alternative is to directly use the knowledge that clients have of other clients. This is done in group lending schemes with joint liability and joint future benefits in case of compliance. The jointness of liability and benefits theoretically provides the incentive to screen clients thoroughly so that group members only accept trustworthy persons as other members of the group. In order to function well, the size of the group has to be small so that members can know and monitor each other. Otherwise, cooperation under their 'assurance game'[7] might rapidly break down. As in many of Nitlapán's local banks, this second alternative can be combined with the first, to create a two-stage selection procedure. Apart from being more effective, any of the alternatives of local client screening and selection also serve the purpose of cutting administrative costs for the bank, by reducing and partially transferring them to the local organization. The organizational burden borne by local clients must, however, rapidly become reasonable, otherwise it will not be worthwhile for them to have access to the financial services.

Rural finance and enforcement costs

Evidently, it is crucial for a viable financial system to enforce a correct execution of the contract terms and to avoid opportunistic 'free-riding' behaviour.[8] In non-conventional financial systems, different alternative enforcement mechanisms can and usually have to be combined. Providing appropriate economic incentives to induce self-enforcement of the financial contract constitutes a first important mechanism. Positive incentives for timely repayment can be created by rewarding compliance with interest rebatements and, more important, access to new benefits in the future (such as higher and/or longer term loans, other services). Positive incentives are usually low-cost to implement and tend to be very effective because of the self-enforcement effect on clients. Given the extreme liquidity crisis of rural producers in Nicaragua, the promise of access to higher and longer term loans in the case of compliance evidently provides a strong incentive to repay. Sequential credit provides for a very effective *short term* incentive for compliance, often resulting in excellent repayment rates (over 90 per cent in Nitlapán). However, it is clear that to the extent that the loans will come closer to the required overall financing need in the future, the progressivity principle could rapidly lose its effectiveness, since further increments in loans and terms are deemed to become less desirable.

Negative economic incentives are provided by the threat of sanctions in the case of non-compliance. A first mechanism is legal sanctions, which usually end up in the sale of collateral if the delinquent client is unwilling or unable to pay. Although legal guarantees and sanctions should not be discarded completely in a non-conventional system, their effectiveness should not be overestimated, because they are problematic to implement effectively. Another important problem of legal sanctions is their excessive cost compared to small loan sums, which rules out legal action as a current mechanism for loan recuperation.[9]

The threat of legal sanctions will have to be combined with a social mechanism to provide for additional and more effective pressure on potential 'free riders'. The existence of joint liability and benefits at village bank or more decentralized group level can substantially improve the effectiveness of the enforcement structures. Joint liability and joint future benefits in the case of collective compliance result in strong peer pressure on the potential 'free riders'. The jointness of positive incentives socially strengthens the incentives for individual self-enforcement. Again, group pressure can be organized in a horizontal or a vertical framework. Within the vertical framework, the threat of material or moral sanctions by powerful local leaders can clearly provide an incentive to pay. In this respect, religious leaders or other respected people can also play an important role by providing a kind of moral guarantee for repayment. People will indeed think twice before risking their reputation as a 'good Christian' or as a 'trustworthy villager'. The sanctions within the horizontal framework are of a similar nature but they are imposed by equals with whom a close personal relationship exists. Unless the potential free riders allow for a radical exit option,[10] these sanctions can be expected to be very effective and to have a high degree of internal legitimacy.

The local social framework around the financial system should in the medium term also provide a substitute for the short term positive incentives as a means to encourage compliance. To the extent that the local society gradually appropriates the local financial system as a part of their natural environment,[11] compliance with the financial system can increasingly become a social habit and ultimately even be seen as a moral obligation. Such internalization of the rules of the game represents the local institutionalization of the financial system. Given the permanent ineffectiveness of abstract legal sanctions, this local institutionalization is a prerequisite for the sustainability of the financial system, especially for keeping costs under control for both bank and clients. It is in this sense that 'the process of institutionalization is an essential aspect of development' (Uphoff, 1993, p. 615).

How can we envisage the social 'magic' of institutionalization?[12] Uphoff states that 'institutionalization is a process, and organizations can become more or less 'institutional' over time to the extent that they enjoy special status and legitimacy for having satisfied people's needs and for having met their normative expectations over time' (Uphoff, 1993, p. 614). This view on the institutionalization process can be made to accommodate the rational choice 'repeated games' view of sustaining a cooperative order on the basis of individual interest. Following Fafchamps (1992, p. 150) a strategic profile for an institutionalization process could 'be constructed specifying-a cooperative path and minimax punishments for each participant.' Mutual individual benefits from cooperation and sufficiently low-cost credible sanctions are definitively important conditions for the prospects of successful institutionalization. Falchamps, however, also indicates that the process always operates as a social network of personal ties, resulting in 'a mesh of interpersonal relationships'. 'Lineage, kinship, neighbourhood, or consanguinity often are major axes of solidarity networks, but friendship and patron–client relationships also matter.' (Fafchamps, 1992, p. 158.) This addition enables an excessively individualistic rational choice view of the process of institutionalization to be avoided. The logic of individual interest through cooperation is supplemented by and unavoidably linked to social processes that form and sustain convergent expectations that gradually transform themselves into social habits and, ultimately, into moral norms.

Financial institution building for development in the Nicaraguan institutional context

Above, we identified the main theoretical principles of a non-conventional organizational framework for rural finance. Three components are crucial: some form of local client selection, incentive mechanisms to enhance individual and/or collective self-enforcement and a social pressure mechanism that should eventually evolve into institutionalized sanctions. Operationally, several alternatives remain open with respect to the precise characteristics of these components. Different mixes of vertical and/or horizontal mechanisms can be chosen, and the relation of bank-client interface with the local structures can be envisaged in different ways. Depending on the specific nature of local institutional environments, different organizational arrangements can and should be set up to buttress the financial transactions. In the Nicaraguan institutional environment the main issue is to articulate the financial system with the different shattered vertical governance structures at the village level.

One possible strategy to deal with this environment is to ignore existing governance structures and to create a minimal local bank–client interface with artificially created social control mechanisms. Such would be the case in systems that relate directly to a simple structure of small joint liability groups without any other intermediation.[13] Such systems do not structurally articulate with the dominant local social networks and are thereby able to avoid many complicated power struggles within the financial system. From a bank management point of view they are thus easier to handle than systems that involve clients more actively. In the short term, repayment can be secured by a combination of sequential credit with joint liability and benefits. In the longer run, it is less clear how repayment can be guaranteed without some kind of identification of the clients with 'their' bank. This need of identification constitutes a bottom limit to the 'minimality' of these systems, unless we assume a situation where compliance with the law and/or one's word is very high – a hypothesis not so appropriate in Nicaragua.

Another problem with the 'minimal' option is that it articulates poorly with the local development dynamics. The organizational framework is limited to an artificial structure of relatively isolated small solidarity groups. Often a predominantly financial perspective combined with a social concern defines the strategy. This translates into a concentration on the poorer social sectors and into a preference for small, short-term loans. This has the advantage of avoiding strong interest of the powerful in the territory and thereby guarantees local 'power compatibility'. A positive characteristic of the 'minimal' system is certainly the horizontal nature of relations within the groups.[14] When functioning properly, the joint liability and benefit groups can provide for a strong and self-reinforcing mechanism of 'social capital accumulation': building mutual confidence shared norms and cooperation for problem solving between equals. Its positive impact will, however, remain limited if this dynamic is confined to the groups and not linked with the broader institutional framework of the village. Ultimately, a narrow focus on financial institution building might turn out to be self-defeating if clients are left to market whims to solve other commercial, technical, legal and organizational bottlenecks.

Therefore, a strategy more explicitly relating to the existing local governance structures might offer more promising prospects in terms of developmental impact. Such a strategy, however, faces the challenge of coping with the inherited vertical power structure that is often inimical to an effective management of the financial system. In the Nicaraguan context of extreme shortage of liquidity, the high benefits for villagers deriving from a sustainable credit offer can, however, be used to enforce more transparent local governance structures as a precondition for access to finance. Moreover, institution building for viable non-conventional rural finance could become an important means of promoting a more appropriate, more horizontal local institutional rearticulation with better prospects for tackling development constraints. The Nitlapán experiment offers some lessons and possibly some elements of a model.

From the outset, the Nitlapán banking system had a development perspective. In this, the need for active involvement of local client organizations in the management and policy of the system was affirmed and the system was structured as a network of relatively autonomous 'village' banks. When the initiative started in 1989, the strategy consisted of promoting and supporting local governing bodies based on the so-called 'natural leaders', i.e. the local *caciques*. Local bank directorates were formed to administer, select and monitor loans. The process of creation of these bodies was largely left to the territories. This resulted in directorates that represented just one of the dominant social networks in the communities. Despite a deliberate option to the contrary from Nitlapán, the spontaneous local processes excluded the possibility of embracing various social networks in the organization. Some of the local banks functioned properly under the vertical control of responsible leaders who, given the sequential credit arrangement, wanted to safeguard future access to credit for themselves and their 'constituencies'. Many banks, however, experienced problems that correspond closely to the theoretical disadvantages of vertical governance structures. In one coffee cooperative, distrust of the local leaders was so great that hardly any clients repaid their debts despite the benefits pending from compliance. In the other cases, lack of transparency and the 'natural leaders' violation of the rules for personal benefits and favouritism[15] contributed to discontent among clients, lagging repayment and ongoing tensions both within the local organization and with Nitlapán. Potential clients not belonging to the dominant network were excluded everywhere.

A new and more viable model of organization is gradually emerging. One of the crucial components of the new model is the obligatory and controlled involvement of all dominant social networks. The new bank directorates now represent the various, often polarized social networks of the local communities. The acute need for finance as well as a growing awareness of shared local interests provides the incentive to accept the imposed condition of local territorial cooperation. Also the rules of the financial system have been changed considerably. Unwarranted romanticism about local autonomy has given way to a much more active and interventionist role of Nitlapán as an external facilitator of local cooperation. A number of detailed rules of the game are laid down in a periodic contract between the national network and the local bank. These comprise a pre-established agreement on the scheduled increase in credit for the next period and the terms of repayment for the current period. The contract also establishes the principles of local banking procedures and policy, such as the mandate of the directorate, the selection committee and the credit promoter of Nitlapán, general criteria for eligibility of clients (group formation, economic activities), standard rules for accounting and administration, and so on. The periodic contract functions as a system of internal laws with which everyone, including those who exercise power, has to comply. It is the role of the external promoter and recently appointed local inspectors to guarantee compliance with the rules and serve as powers countervail to local leaders who might attempt to abuse the system. The presence of competing social networks also functions as an internal control, since the representatives tend to control each other. Respect for the rules supports the institutional space in which the competing social networks cooperate.

The model emerging in the Nitlapán network can be described as a kind of local 'constitutional' order[16] with a variety of checks and balances on the leaders, both at the level of the local organization and at the bank-client interface. This strongly reduces the possibilities of clientelistic favouritism, while it also protects local leaders from the pressures that can be put on them by their 'constituency' to obtain benefits beyond finance. This allows a transition towards more transparent 'civic' relations from within the prevailing vertical social structures. This could be a promising model for promoting a transition towards more horizontal and 'civic' local rural structures. Increasing positive experiences of civic cooperation for finance have the potential to create a platform for effective and more horizontal cooperation in other fields. The need to link the local development bank with local economic development provides for an additional incentive in this direction. In this way, alternative finance could indeed become a 'principal means of reorganization of local cooperation' (Nitlapán, 1995, p. 30).

5 CONCLUSION

After more than a decade of profound structural reforms, rural Nicaragua is searching for an appropriate institutional rearticulation. One of the main problems is the relative incompatibility between the vertical clientelistic institutional inheritance and the objective equality of the land property structure. A transition towards more horizontal governance structures would therefore be preferable. Our analysis linked the challenge of organizing viable rural finance with the institutional deficiencies of rural Nicaragua. We identified the theoretical components of viable institutional arrangements for non-conventional rural finance and concluded that such financial institution building could provide an effective means to promote more horizontal civic rural institutions. Such new institutions are absolutely necessary for consolidating the equitable rural development path that corresponds to the present land property structure.

Notes

1. Much of this contribution has been inspired by development experiences of Nitlapán, a research and development institute of the Universidad Centroamericana of Managua. I wish therefore to express my indebtedness to the institute's collaborators, and particularly to Concepción Almanza, Carlos Barrios, Elisabeth Campos, Eberth Hernandez, Oscar Manzanares, René Mendoza, Alfredo Ruíz, Eva Sanchez and María de los Angeles Solarí, who have all contributed to shape my ideas. Helpful comments were also made by Peter Marchetti, Tom De Herdt, Stefaan Marysse and Ruerd Ruben. The views expressed remain my own responsibility and are not necessarily those of Nitlapán.

2. It is evident that in 1981 the 'contras' originated as a US-sponsored proxy army based on the remnants of the Somozista Guardia Nacional, but in the field this army was gradually appropriated by the poorer peasant strata of the agricultural frontier (Bendaña, 1990).

3. In this respect, it can be noted that due to greater exigencies of international donors many foreign-sponsored rural development projects are hardly being executed at present due to a lack of capacity to get the capital to the rural target groups and to guarantee its productive use. Nitlapán estimated that without fundamental institutional changes only about 4.2 million US$ of the 25.9 million US$ that were programmed for the period 1995–7 for rural development projects would be still available by the end of that period. They concluded that 'it is not acceptable to continue to ask for new disbursements of foreign aid without the correction of the errors of the past' (Nitlapán, 1994, p. 25).

4. In this sense, the creation of a viable and democratic market society in the Nicaraguan countryside requires both the destruction of inherited cultural forms (with their positive and negative characteristics) and the creation of

new forms that are beneficial for stable cooperation between *a priori* more autonomous individuals and families. This corresponds to Hirschman's conclusion about the possible rival interpretations of market society (Hirschman, 1982, p. 1483): its emergence implies both the destruction of existing society (negative interpretation) and the creation of a new culture that sustains the market order (positive interpretation).

5. Unless indicated otherwise, the statistical data for this section come from Nitlapán (1994b, 1994c).

6. The survey covered the departments of Estelí, Madriz, Nueva Segovia, Matagalpa, Jinotega, Boaca and Chontales.

7. In the joint liability group scheme, net gains for all players in the case of compliance by all are superior to all other outcomes. However, the net gain in the case of compliance when the others defect are lower than when all defect. Without sufficient confidence and mutual monitoring, each player will thus be reluctant to comply and defection by all might be the logical outcome. Effective coordination to reach the optimum is, however, easier than in the prisoner's dilemma game because gains from defection are never higher than under the optimum. Free riding therefore does not pay. The issue that remains is the 'sucker' problem. For an interesting and more developed game theory approach to group lending, see Besley and Coate (1985).

8. Given the rather chaotic nature of recent history in Nicaragua, and more particularly the severe erosion of the concept of credit itself, the danger of free-riding is possibly more severe in Nicaragua than in other countries.

9. Occasional legal action can, however, have an important psychological effect. In one instance, a local bank of Nitlapán obtained the right to publicly sell a television set and a bicycle to recuperate a loan. The cost of the legal procedure was hardly covered, but the act did have a significant impact in the perception of the clients.

10. In the interior regions it turns out to be very risky to provide credit to the landless wage-labourers (*colonos*) for this reason. Many of them change their residence often and some even decide to change their luck in the agricultural pioneer frontier, using their loan as starting capital.

11. Such local appropriation can also be enhanced by requiring clients to save in the bank or even, as in Nitlapán's case, to oblige clients to become shareholders of their bank for a stipulated minimum amount of capital.

12. The process can be termed social 'magic' because the logic of the rational individual actor will never produce the required outcome in terms of moral obligations. Rational choices alone cannot overcome distrust and will thus logically lead to free riding and opportunism (Platteau, 1994, pp. 757–64).

13. This seems to be the option in a new unconventional financial system of the cooperative branch of the peasant union UNAG, FENACOOP (Oral communication from D. Pommier).

14. Care needs to be taken, however. In practice, the joint liability group might not always correspond to the 'ideal type' of equals joining equals. In the Nitlapán banks in the interior, many of the groups turned out to have been formed by one accommodated producer joining a number of poor dependent land labourers or peasants. These groups closely resembled the clientelistic

practice of providing bail. One can also expect a high prevalence of family-based groups that may not be very horizontal either.

15. This emerged in a number of committees with completely non-transparent procedures where local directors attributed themselves and their relatives 'illegally high' loans and/or 'forgot' to pay in time. A recent rule that bank 'directors' could not be elected if they were behind in their repayments prevented several directors from being re-elected.

16. The concept is taken from Sabel (1993). He presents the 'constitutional order' as a governance structure that differs from the two alternatives identified by Williamson: i.e. vertical authoritarian structures and horizontal market autonomy (Williamson, 1993, pp. 16–17).

Bibliography

ASODERV (1993) *Investigación sobre el campesinado de la V región. Su idiosincracia y sus valorizaciones políticas y organizativas* (Managua: ASODERV).

Barraclough, S. and P. Marchetti (1985) 'Agrarian Transformations and Food Security in the Caribbean Basin', in G. Irvin and X. Gorostiaga (eds) *Towards an Alternative for Central America and the Caribbean* (London, Allen and Unwin).

Bastiaensen, J. (1991) *Peasants and Economic Development: A case-study on Nicaragua*, Ph.D. thesis in Applied Economics, University of Antwerp, UFSIA, Faculty of Applied Economics.

Bastiaensen, J. (1994) 'Una Vía Campesina en Nicaragua: Una Reflexión de Cara a los Desafíos', in J. De Groot and M. Spoor (eds) *Ajuste Estructural y Economía Campesina*, Managua, ESECA-UNAN.

Bendaña, A. (1991) Una Tragedia Campesina: testimonios de la resistencia, Managua, Edit-Arte, CEI.

Besley, T. and S. Coate (1995) 'Group Lending, Repayment Incentives and Social Collateral', *Journal of Development Studies*, Vol. 46, pp. 1–18.

CIERA (1989) 'Matiguas: campesinado y formas organizativas', *La Reforma Agraria en Nicaragua: 1979–1989*, Vol. IV. Economía Campesina CIERA, pp. 275–316.

De Franco, M.A. and R.J. Sevilla (1994) *La Economía Política de la Ayuda Externa en Nicaragua: Finanzas Públicas, Desarrollo Humano y Crecimiento Económico*, Managua, Nitlapán-CRIES.

De Herdt, T. (1995a) *Behind Platteau: A critique of Platteau's critique of Granovetter's critique of the concepts of trust and malfeasance in the New Institutional Economics*, Antwerpen, University of Antwerp, UFSIA, Center for Development Studies, mimeo.

Doligez, F. (1994) 'Estrategía institutional para el Financiamiento Rural en Nicaragua', Misión de Apoyo a la red de Bancos Locales de Nitlapán-UCA, Managua, IRAM-Nitlapán.

Fafchamps, M. (1992) 'Solidarity Networks in Preindustrial Societies: Rational Peasants with a Moral Economy', *Economic Development and Cultural Change*, The University of Chicago Press, pp. 147–74.

Gambetta, D. (1988) 'Can we trust trust?', in G. Gambetta *Trust: making and breaking cooperative relations* (London: Basil Blackwell).

Hirschman, A.O. (1982) 'Rival Interpretations of Market Society: Civilizing, Destructive, or Feeble?', *Journal of Economic Literature*, Vol. XX, pp. 1463–84.

Hoff, K., A. Braverman and J. Stiglitz. 'Introduction' in *The Economics of Rural Organization*, pp. 1–29.

Houtart, F. & G. Lemercinier (1992) *El campesino vomo actor. Sociología de una comarca de Nicaragua, El Comejen*, Managua, Ed. Nicarao.

INRA-OEA (1992) *Marco Estrategico de la Reforma Agraria. Balance Preliminar sobre la oferta y demanda de tierras para la formulación de una estratégia de reforma agraria*, Managua.

Kaimowitz, D. (1987) *Agrarian Structure in Nicaragua and its Implications for Policies towards the Rural Poor*, Ph.D. dissertation, The University of Wisconsin, Madison.

Kaimowitz, D. (1989) 'La Planificación Agropecuaria en Nicaragua: de un Proceso de Acumulación basada en el Estado a la Alianza Estrategica con el Campesinado' in R. Ruben and J. De Groot (eds) *El Debate sobre la Reforma Agraria en Nicaragua, Managua*, ECS/INIES.

Lipton, M. (1993) 'Land Reform as Commenced Business: the Evidence Against Stopping', *World Development*, Vol. 21, n° 4, pp. 641–57.

Maldidier, C. *et al.* (1995) *El Campesino-Finquero y el Potencial Económico del Campesinado Nicaraguense*, Managua, AID, Nitlapán-UCA, DDW-UFSIA, (forthcoming).

Marchetti, P. (1994) *Experimentación con Nuevas Modalidades de la Educación Popular para el Desarrollo Local*, Managua, Nitlapán-Universidad Centroaméricana, mimeo.

Nitlapán Equipo de Investigación Sectorial Rural (1995) 'Informe Ejecutivo Preliminar. Encuesta Rural para el Proyecto de a Agraria y Ordenamiento de la Propiedad Agraria', Managua, 15 de Agosto 1995, mimeo.

Nitlapán (1994a) 'Financial Services Program', Managua, Nitlapán-UCA, mimeo.

Nitlapán (1994b & 1994c) 'Situación y Perspectivas de las Neuvas Estructuras Institutionales de Financia-miento Rural', Informe Final & Resumen, Estudio preparado a solicitud de la Autoridad Sueca para el Desarrollo Internacional, Managua, Nitlapán, Universidad Centroamericana.

North, D.C. (1990) *Institutions, institutional change and economic performance*, Cambridge USA: Cambridge University Press.

Platteau, J.-Ph. (1994) 'Behind the market stage where real societies exist – Part I: The Role of Public and Private Order Institutions – Part II: The Role of Moral Norms', *The Journal of Development Studies*, Vol. 30, No. 3–4, pp. 533–77 & pp. 753–817.

Putnam, R.D. (1994) 'Democracy, Development, and the Civic Community: Evidence from an Italian Experiment', *Culture and Development*, The World Bank, pp. 33–73.

Sabel, C.F. (1993) 'Constitutional ordering in historical context' in F. Scharpf (ed.), *Games in Hierarchies and Networks; analytical and empirical approaches to the study of governance institutions* (Frankfurt/Colorado: Campus Verlag/ Westview Press).

Thorbeke, E. (1993) 'Impact of State and Civil Institutions on the Operation of Rural Market and Nonmarket Configurations', *World Development*, Vol. 21, No. 4, pp. 591–605.

Uphoff, N. (1993) 'Grassroots Organizations and NGOs in Rural Development: Opportunities with Diminishing States and Expanding Markets', *World Development*, Vol. 21, No. 4, pp. 607–22.

Williamson, O.E. (1993) 'The Economic Analysis of Institutions and Organizations in General and with Respect to Country Studies', Economics Department Working Papers No. 133. Paris, OECD.

13 Rural Lending by Projects: Another Cycle of Unsustainable Interventions in Credit Markets? An Analysis of Case Studies in Central America

Harry Clemens and Cor Wattel

1 THE DEBATE ON THE SUSTAINABILITY OF RURAL CREDIT PROGRAMMES

The sustainability of rural credit programmes arose as a worldwide theme of debate in the 1970s. An emerging body of literature severely criticized the state intervention in rural financial markets, arguing that these interventions hampered financial development and did not reach their objectives of broadening access to credit to the poorer segments of the rural population (for example Adams *et al.*, 1984; Von Pischke, 1991).

The critique was directed mainly against the performance of the state-owned rural development banks which had been created mainly in the sixties, as part of a broader policy of integrated interventions in rural markets for land, credit, output and inputs. The creation of the rural development banks served two main purposes. Firstly they were designed to help integrate the Central American economies more effectively in the world economy. To this end, considerable flows of capital were provided to promising agricultural exports (*e.g.* coffee, cotton, cattle). Secondly, development banks were intended to broaden the access to credit for small producers in the rural economy, under the assumption that access to the means of production is the determining factor in underdevelopment.

The critique on the development banks basically drew on two arguments (Braverman and Guasch, 1993, pp. 54–7): their regressive distributional effects and their lack of financial sustainability. Unlike the objective of the development banks to broaden the access to credit, in several cases the *distributional effects* were proven to be regressive. The distributional effects arise mainly because of the subsidy element in the credit provision by development banks, which induces an implicit income transfer to the borrowers. This income transfer is larger for borrowers who succeed in obtaining larger loans. One of the reasons for this credit allocation bias in favour of the larger holdings is a cost-benefit argument for the lender: it is cheaper for a bank to allocate credit to a few large holdings than to many smallholders. There are also socio-political causes, as the large landowners have a better capacity to lobby for credit programmes which serve their needs, and thus obtain larger implicit subsidies.

The other main argument against the performance of the rural development banks is their *lack of financial sustainability*. The interest rates were usually set at low levels, sometimes even negative in real terms, because of the imposition of interest rate ceilings. This provided the banks with too little income to cover operating costs. It also obliged them to set the deposit rates at a low level, which hampered the mobilization of local savings. Financial unsustainability is also generated by the high default rate that characterized many of the development banks' portfolios, caused by deficiencies in the screening, supervision and enforcement procedures. One of the underlying factors is the reliance of these banks on external funding, with a low priority for the recovery of the loans. A third factor thwarting the sustainability of many development banks is the high level of their administrative costs, partly due to the emphasis on small-scale lending.

Following all the critique on the performance of the rural development banks, financial reforms are being implemented in all Central American countries. The reforms generally aim at the sound development of the financial sector itself, and include several aspects: the restructuring of the development banks, a greater independence of the central banks *vis-à-vis* the governments in setting monetary policies, and a stronger supervision of the private banking system by the monetary authority.

The restructuring process in the development banks tends to give high priority to their financial and institutional sustainability, and less importance to the original purpose of broadening the access to credit. This brings us back to the situation where the problem of the access to formal rural credit remains an unresolved question.

Parallel to the growing critique on the role of the development banks, an increasing number of non-bank credit programmes has been established.

These credit programmes have primarily pursued the objective of broadening the access to credit to small rural producers and to promote their integration in the local markets; often they are intimately related to the provision of other non-credit services, such as technical assistance, training, organizational support. In a certain sense, their objectives are similar to the objectives of the development banks, albeit that their scope is more limited to target groups in specific regions or sectors.

As yet there has been relatively little analysis of smaller-scale credit programmes of a non-bank character. The discussion on rural financial markets tended to focus on the role of the development banks and the impact of subsidized credit facilities on financial development at national level. The scope of the non-bank programmes is generally more restricted, but at a local level can have an impact of equal importance on the credit availability.

One could ask whether these smaller-scale interventions have generated a more positive impact than the larger-scale establishment of specialized development banks. They would do so if they succeeded in progressively redistributing income and access to credit, and in maintaining financial sustainability.

In the following sections, we will comment a number of non-conventional credit programmes in Central America of a very diverse nature, examined throughout the last two years (1994–5). These cases will provide some insights on whether the non-bank lending programmes offer a viable alternative for reaching the small-scale producers in rural areas, or merely repeat the same errors made previously by the development banks.

2   BASIC FEATURES OF CASE STUDY CREDIT PROGRAMMES

Non-conventional credit programmes aspire to fill the gap between commercial bank lending and informal credit, through the development of financial technologies which satisfy the demand for financial services of households that do not have access to credit and saving facilities of traditional banking institutions without incurring high costs associated with informal credit. Most programmes concentrate on credit services, but some include saving facilities as well. In this paper non-conventional credit is understood as all credit provided by special programmes designed to broaden the access to institutional credit. These programmes may use administrative services of banks (trust funds), or be administered by some other kind of institution (NGOs, cooperatives).

There are no reliable statistics on the importance of non-conventional credit for the region. However, the share of this type of credit in rural lending appears to be substantial, especially in Nicaragua and Honduras. In Nicaragua it can be estimated that non-conventional programmes served more than 38 000 clients in 1993, while commercial banks (excluding trust funds administered for Integrated Rural Development projects) served only some 31 000 clients (BCN, 1994; Nitlapán, 1994). However, average loans from non-conventional programmes are much smaller than those from commercial banks and merchant firms; it is estimated that in terms of disbursements non-conventional credit contributed to 8–10 per cent of total rural finance. In Honduras a survey in three regions found that in 1994 the number of clients served by NGOs was similar to that served by the National Development Bank or private commercial banks (González Vega and Torrico, 1995). In El Salvador a national survey found much less evidence of the importance of non-conventional financing in early 1991 (Cuevas *et al.*, 1991), though this type of financing probably grew in importance after peace treaties were signed in 1992. According to the 1993 Land Tenure Survey in El Salvador, 6.1 per cent of the farmers used NGO credit during the 1992/93 agricultural cycle, compared to 10.3 per cent for formal credit (Benito, 1993; Quintanilla, 1994).

Our analysis is based on a sample of 19 case studies of credit programmes in Central America supported by foreign donors, including multilateral and bilateral governmental agencies, and non-governmental organizations (NGOs). Primary information was collected and analysed by researchers from the Rural Development Consult Centre of the Free University of Amsterdam during 1994 and 1995 (in the case of three cases supported by information collected by the Regional Unit of Technical Assistance project from the World Bank), with the exception of the FINCA programme for which information on 1992 from a secondary source was used (González Vega, 1993). Most programmes had between two and five years' experience with credit operations.

The study includes 19 cases only, but the diversity of cases included can be considered representative of rural credit market interventions by projects in Central America during the early nineties. Five rural development programmes with a credit component are included, plus three specialized credit projects, six credit programmes of NGOs, one case of a multiple services cooperative, two cases of credit programmes administered by a federation of cooperatives and two cases of credit programmes channelled through marketing cooperatives.

The case studies are rather diverse in scale of operations, mode of operation, funding or external executing agency. Only a few programmes include mobilization of savings as part of their development of financial services; most provide credit services only. All programmes target their funds to small producers or poor rural dwellers; that is excluding large farmers from their potential clientele. However, different segments of the target group are served: some provide services to individual farmers, some to cooperative organizations, and some to other groups of rural dwellers. In two programmes the organization of communal banks was stimulated. The scale of operations varies from US$ 64 000 to US$ 6 200 000 (see Table 13.1).

One of the striking aspects of non-conventional rural credit programmes is the variety in operational arrangements. Some provide loans to groups of borrowers only, in order to reduce transaction costs, other provide individual loans only, arguing that group responsibility gives rise to free riding. Some try to bridge the gap between target groups and financial institutions by providing trust funds to the latter in combination with other interventions directed at improving the creditworthiness of clients. Others try to develop new financial institutions, with or without a participation of the target group. Some programmes develop specialized financial services, others make use of interlinked contracts with marketing. With respect to collateral, some try to rely on social collateral (group lending; personal guarantors), others experiment with different kinds of *prendas* or a combination of real and social collateral. In the next section we will analyse different operational arrangements aimed at optimal screening, monitoring and enforcement and low transaction cost.

3   ACHIEVEMENTS AND SUSTAINABILITY OF NON-CONVENTIONAL RURAL LENDING BY PROJECTS

In this section we analyse the 19 cases on three issues. The first is the question of whether there is evidence of regressive distributional effects as a result of the credit provision. The second is whether the programmes have succeeded in developing institutional set-ups that promote innovations in the screening, supervision and enforcement procedures and provide an organizational structure with adequate incentives. The third issue is to what extent the programmes studied have achieved financial sustainability.

Access by target groups and effects on income distribution

One of the criticisms of national development banks is that most of their interventions have proven to be regressive (Braverman and Guasch, 1993, pp. 54–5). Typically, ceilings that are below market rates are imposed on

Table 13.1 Basic features of credit programme case studies

Program/ Institution	Country	Credit portfolio or fund (US$)	Admin. agency (1)	Origin of funds	Target group
NGO 1	Honduras	64 000	Own	Grant by foreign NGO	Land reform groups
NGO 2	Honduras	< 70 000	Own	Grant by foreign NGO	Peasant shops
NGO 3	Honduras	80 000	Own	Loan from FHIS	Non-agric. micro-enterpr.
RDP 1	Honduras	123 000	C&CS	Grant by foreign Gov.	Agric. cooperatives
NGO 4	Honduras	150 000	Own	Grant by foreign NGO	Community banks/ micro enterprises
COOP 1	El Salvador	150 000	Own	Grant by foreign NGO	Fishery coops.
COOP 2	Nicaragua	200 000	Own	Mainly concessional loans	Agr. coops./ groups
COOP 3	El Salvador	250 000	Own	Grant by foreign NGO	Coffee producers coop.
COOP 4	Nicaragua	257 000	Own	Grant by foreign NGO	Agric. market- ing coop.
NGO 5	Guatemala	280 000	Own	Grant by foreign NGO	Groups of small farmers
CRP 1	Costa Rica	700 000	Own	Grant by foreign Gov.	Small farmers/ rural enterpr.
CRP 2 (2)	Costa Rica	700 000	Own	Grants/long term loans	Rural commu- nity banks
CRP 3	El Salvador	775 000	Own	Grants by foreign Gov.	Agric. produ- cers coop.
NGO 6	El Salvador	> 1 000 000	Own	Mainly con- cessional loans	Small farmers
COOP 5	El Salvador	1 540 000	Own	Grants by foreign NGOs	Agric. produ- cers coop.
RDP 2	Costa Rica	1 700 000	Bank	Grant by foreign Gov.	Small farmers
RDP 3	Guatemala	4 477 000	Bank	Concessional loan IFAD	Very small farmers
RDP 4	Guatemala	5 200 000	Bank	Concessional loan IFAD	Small farmers
RDP 5	Costa Rica	6 200 000	Bank	Loans IFAD/ BCIE	Small farmers

(1) Administrative agency is either Own staff of project or institution, Credit and Savings Cooperative, or Bank.
(2) In this case data are for 1992.

interest rates inducing well organized groups of large landholders to lobby to gain access.

Though most credit programmes in the sample charge interest rates even lower than the national development banks, there is no indication that loan allocation is diverted to large landholders or other wealthier groups which do not pertain to the declared target groups of the programmes. The result that loans are actually given to the declared target groups can be explained by three factors. First, the relative smallness of the loans of most programmes makes it less attractive for wealthier groups to gain access to these programmes. Second, in most cases the project staff of donor organizations monitor this aspect fairly strictly. In screening clients, many programmes make use of information provided by extension workers who provide technical assistance to these clients. Projects are usually evaluated on numbers of clients (beneficiaries), so project staff are inclined to ensure that credit is distributed to a large number of households. Third, private organizations executing non-conventional credit programmes operate more independently from political influences or pressure from large producers, compared to agricultural development banks.

The declared target group of most credit programmes are small producers. Consequently, most such programmes do not pretend to reach the very poor rural population who lack access to land and capital. However, a few programmes do target their funds on the very poor. Two of them work through community banks: one through peasant shops and another through a combination of promotion of financial intermediation by groups with administration of funds by a national development bank. The programmes which finance the establishment of community banks not only provide credit services but also savings facilities. These programmes, as well as the one which allocates credit to peasant shops, have adapted their supply of financial services to the needs of the target population.

It has been argued that different groups of poor households have different needs for financial services. According to Hulme (1995) the very poor are more concerned with reducing the vulnerability of their livelihood (protective function of financial services; that is consumption smoothing) than with raising income (promotional function). Most non-conventional credit programmes are primarily concerned with the promotional function of credit, with, as mentioned, some exceptions. In fact, the legal framework in Central American countries does not allow for mobilizing savings deposits from the general public, as this function is reserved for supervised financial institutions. Promotion of savings by non-conventional programmes is only possible on a mutual basis, or under the umbrella of a supervised institution.

It can be concluded that the net effects on income distribution from non-conventional rural lending by projects are not regressive like those from former interventions brought about by the creation of national agricultural development banks. Loans are allocated to the declared target groups. These groups mainly include small producers, rather than the rural very poor.

Institutional design

One of the weak aspects of the development bank experience is that the screening, supervision and enforcement procedures proved to be deficient, and this was reflected in high default rates. As a reaction, in several cases the agricultural development banks shifted to stricter collateral mechanisms, which, inevitably, proved to exclude the small-scale producers. The question in this section is whether the non-conventional credit programmes succeeded in designing screening procedures that are more effective in ensuring repayment and more efficient in transaction costs, while not excluding small-scale farmers.

The *screening* function is intended to distinguish ex-ante between good loans and bad loans. In all the schemes studied, *knowledge at the local level* was used in order to make a better appraisal of the quality of clients and projects. Following Bastiaensen (1995), we can distinguish two types of involvement of local knowledge in the screening process: vertical and horizontal. The vertical mechanism implies the *involvement of local leaders* in the selection and approval process. These leaders may be members of religious organizations (priests, church-related or religiously inspired NGOs), agricultural unions or cooperatives, or local interest groups. Local leaders have more information available about the creditworthiness of the potential borrowers, and as such are in a better position to judge the feasibility of a loan. This method is normally used in the community bank programmes, where the board decides on the approval of loans. Several NGO credit programmes and the cooperative programmes also make use of local leaders. Nonetheless, the effectiveness of the involvement of local leaders in screening depends heavily on their motives and on proper incentives. If the leadership is based upon their capacity to make responsible decisions and their abilities in strategic planning, involving leaders in the screening process may be a valuable tool.

It is then important that proper incentives are designed in order to make the leaders' own interests compatible with the interests of the lenders. Basically these incentives should ensure that the leaders have something at stake in case of default. For instance the leader may put his own assets at risk as (partial) collateral for loans recommended by him; or he may sign the credit contract as co-debtor.[1] Other more positive incentives may include remuneration for the satisfactory performance of the loan portfolio, by paying a fee for good performance. Where this type of incentive does not exist, the role of local leadership in screening is only effective as long as the moral inspiration of the leaders outweighs the benefits of allocating bad loans to good friends.

The horizontal mechanism of local screening refers to schemes of *joint liability*. In these schemes, the borrowers provide a personal guarantee for each other's loans, and have to pay if one of the group members defaults. In our case studies, the effectiveness of the joint liability system appeared to depend upon various factors. One is the size of the group: the smaller the group, the more effective the joint liability. In a small group (5–8 persons), mutual confidence and social control is much more effective. Moreover there is more at stake for each individual if one of the group members defaults (see the article on peer monitoring by Stiglitz, 1990). In one of the organizations, for example, the group size is 20–50 persons, which makes it impossible for anyone to monitor the performance of all his group members. This is why FINCA Costa Rica has suspended the system of joint liability in favour of a system of individual guarantors (1–2 per loan).[2]

Another danger for joint liability schemes is the covariance of risk. When all borrowers invest in the same crop at the same time, or invest in a collective project, the capacity to pay of each individual will be equally affected by a fall in profitability.

A substitute for the utilization of local knowledge is the *involvement of agricultural extension officers* in the loan appraisal process. This method is frequently used in larger integrated rural development programmes, and in smaller programmes with a similar strategy. Yet the experience with extension officers screening clients is not very promising. Although the extension officers do build up local knowledge about potential clients and can monitor their performance, they lack the moral authority that is needed to make potential cheaters change their behaviour. There is also an aspect of conflict of interest: extension officers are often so closely involved in the promotion of innovative technologies or crops, that they have an own interest in having these projects financed. They then convert themselves into allies of the clients instead of critical bankers. The second disadvantage of screening by extension officers is that it does not contribute to the sustainability of the credit system: the wages of the extension officers are difficult to sustain in an autonomous credit system.

Another screening mechanism used by the non-conventional programmes is the *interlinkage* of credit with trade transactions. In several cases, marketing is the core business of the organization and the credit activity is directly related to it. The loans are given as advance payments on the delivery of output, or in the reverse case inputs are delivered on credit. The interlinkage functions as a screening method because the trade relationship between borrowers and lenders provides the latter with additional information on the repayment capacity, which creates trust as a basis for a credit relationship.

Interlinkage of credit with operations in other markets is often mentioned as an effective screening and enforcement mechanism. In our sample of non-conventional credit schemes, interlinkages were only found with marketing of output or inputs.[3] The cases include three examples of cooperative trading organizations in which this type of interlinkage operates rather successfully; although a few critical remarks may be made:

- The interlinkage is only secure if the marketing cooperative performs well. It is always possible that the borrowers will decide to sell their product to another marketing channel and thus avoid repaying the loan to their cooperative. The probability of this happening increases when the marketing cooperative offers low prices, pays late or in any other way offers fewer benefits than the competitors. If competition is keen, the effectiveness of trade-credit interlinkages may be eroded.

- The interlinkage alone does not guarantee repayment. In one case, for instance, the procurement contract is complemented with individual collateral. The individual farmers guarantee their loans with a combination of delivery contracts and movable collateral (cows, horses), or in some cases mortgages on land or house. This reinforcement of the collateral requirement proved to be necessary in order to strengthen the personal accountability of each farmer for his own loan. This personal accountability is loosely regulated in many agrarian cooperatives, which has caused the repayment attitude of land reform beneficiaries to deteriorate.

- There are several contractual possibilities of reinforcing the effectiveness of the trade-credit interlinkage. An interesting one, though not applied in the cases under study, is to define the product to be procured (coffee, sesame) not as a future crop to be bought, but rather as an existent product owned by the lender and contractually deposited in hands of the borrower. If the borrower decides to sell his product to a competing company, this is penalized as a crime (robbery), instead of an infringement (reneging).

Without adequate *enforcement procedures*, any screening procedure will remain ineffective. Generally speaking we can distinguish between moral incentives, financial incentives and legal sanctions. Moral incentives refer to the social pressure exerted on the borrowers by their neighbours, which prevents and eventually cures potential default. The financial incentives may include giving the timely paying borrowers access to progressive loans,[4] discounts on interest rates, speeding-up approval procedures, revolving credit lines, etc. Legal sanctions are fundamentally based upon collateral, supported by a legally acceptable documentation of the credit contract.

Moral incentives are used in some way by all of the programmes investigated here. The moral pressure is generally effectuated by the same person or organization responsible for the screening process. The same comments made above on the various screening mechanisms apply to the moral incentive aspects: much depends on the legitimacy of these persons or institutions in the community, and their credibility as a creditor. In some cases the groups of beneficiaries were rather artificial creations with a strongly donor-related identity, and as such did not provide a solid social network that could exert moral pressure on defaulters. Some programmes make effective use of the moral incentives; nonetheless, even in these cases the moral incentives alone do not provide a sufficient enforcement mechanism, but are combined with financial incentives or the possibility of legal sanctions.

Financial incentives are used less often than moral incentives. The most frequently observed mechanism is to deny defaulting borrowers access to new credits. This, however, proves to be a rather unwiddy instrument in practice. The defaulting borrower may not have access to new credits, but can continue to use the existing loan capital to refinance his own activities, without paying interest! Closing access to new credits is therefore only effective if the new credits are more favourable than the old ones (larger sums, continued technical assistance or other project services); in other words, if some progressivity principle is being applied. The progressivity principle, however, is only explicitly applied in the community banks, but not so in other programmes.

The legal aspect is a neglected theme in many of the non-conventional programmes. In a number of cases (mostly NGO) there are deficiencies in the formalization of the credit contracts. Any of these formal deficiencies may make legal enforcement virtually impossible.

The central legal issue, however, is the collateral requirement. Immovable collateral (mortgage on land or house) is seldom required in the non-conventional credit programmes. The main reason for this is that the enforcement of this type of collateral is socially unacceptable, the more so because of the social nature of most of the credit programmes. Apart from this motive, the possession of land and other marketable assets among the target population is rare, and in many cases is not formally registered (land titles). Moreover, the transaction costs of registering and executing mortgages are too high to be justified by small and short-term loans.

Movable collateral (cattle, food stocks) is used by various programmes, but is not effective in all cases. The movable collateral can function relatively well in situations where the lending organization is equipped to deal with agricultural products and sell them if necessary. This is the case for the coffee or sesame trading cooperatives that use the coffee and sesame harvest as a partial collateral for the loans. These cooperatives are also more capable of selling cattle, which may serve as collateral, because their associates are farmers used to buying and selling in agrarian markets. For non-agrarian organizations, such as NGOs or state-managed credit projects, it is generally more difficult to make movable collateral effective, as their organization is not equipped to enforce these guarantees.[5]

Personal guarantees are often used in situations where neither mortgages nor movable collateral are effective. Almost all of the programmes included in our sample applied some type of personal guarantee: joint liability, group liability, one or two guarantors. The effectiveness of this type of guarantee was commented on in the previous section.

Credibility is a separate point about enforcement that deserves to be commented on. The lender's credibility largely depends upon his capacity to successfully recover the loans. Our study included several examples of lending organizations that demonstrated a flexible attitude towards default and were punished by ever increasing default rates. The borrowers tended to identify these organizations as donors instead of lenders. They were no longer willing to pay their loans while others did not.

A final aspect on enforcement is the absence of financial incentives to the credit officers that might stimulate a higher pressure on repayment and screening. None of the lending programmes investigated gave bonuses for low default rates.

Financial sustainability

The financial sustainability of a lending programme can be defined as its capacity to reproduce the cycle of lending and recovering credits without structural subsidies. The financial sustainability is determined by the degree to which the income (interest, fees) covers the costs of lending.

To start with, the vast majority of the investigated programmes charge interest rates below the market rate. Half the programmes charge interest rates far below the market rate, and another third charge rates that are less than three per cent below the market rate. Only two of the programmes sampled charged interest rates above the market level.

There are two comments to be made on this phenomenon. First, it is very improbable that the costs of lending of these non-conventional lending organizations are lower than the costs of bank lending. Small-scale lending is relatively expensive, measured in terms of the costs per peso loaned. This argument would favour small-scale credit programmes charging interest rates higher than the market.

On the other hand, the real interest rates in the market are relatively high. The passive interest rates are pushed upward by the trade and fiscal deficits, and the spread between active and passive rates is kept large because of the low level of competition in the financial markets in Central America.[6] This might imply that non-conventional credit programmes may cover their costs with interest rates lower than those charged by the banks.

The operating costs of most credit programmes are relatively high and cannot always be measured. In many cases the costs of lending are not registered explicitly, as they are mixed with other categories of expenses (training, extension) or are heavily subsidized. This is true for more than half of the programmes in our sample. Only three programmes succeeded in maintaining operational costs lower than 6 per cent of the portfolio, due to the subcontracting of the credit administration to specialized institutions, a combination of loans services with marketing, or because the programme specialized in financial intermediation.

The scale of operations obviously influences an organization's ability to reach the break-even point. Half of the programmes administer funds smaller than US$ 500 000 which is probably too small to cover operating costs, even if professional credit management techniques are applied.

Very few non-conventional programmes attract local savings. Only two of the programmes investigated have some kind of mutual savings facility or obligation, both of them systems of community banks. A few other programmes, mainly of a cooperative character, ask their members to deposit capital shares or retain profits for the same purpose. Nonetheless, the total savings or capital shares stay far behind the size of the loan portfolio, which implies a continued dependence on external funding.

In the next section financial sustainability is interrelated with the screening and enforcement variables.

Determinants of sustainability

A proper institutional design is a necessary prerequisite for financial sustainability. If this is not the case, it is to be expected that deficient screening and enforcement mechanisms will lead to high default rates, and thus financial unsustainability.

For our analysis of financial sustainability we calculated a proxy by estimating losses due to default rates and comparing the interest rates of the programme with market rates. It was assumed that market rates take into account a provision cost of three per cent for non recovery of loans. For market interest rates the interest rate charged by commercial banks on short term rural credit was taken as a proxy (a uniform rate for each country). In 16 programmes the interest rate was below market rates, in one programme both were equal, and in two programmes programme rates were higher. Negative differentials were taken as a loss. So it is assumed that these were not due to lower administrative costs. It might be that programmes have lower costs (for example due to lower supervision costs, less paperwork and larger loans by group lending), but in most cases they are higher. On average, it was assumed that non-conventional programmes incur a three per cent higher administration cost than competitors because, on average, loans are smaller. On the other hand, lower interest rates might be viable in the short term due to access to donations or cheap credit lines for financing loans to rural households. However, in the long term, sustainability of the programme will depend on recovery of market cost of capital.

Default rates ranged from 0 to 60 per cent. For most cases the estimated loss due to non recovery was taken as one third of the percentage of credit portfolio in arrears. In some cases, where available information indicated a difference between arrears and a reasonable expectation of repayment, another percentage was taken.

The programmes analysed were classified into four groups according to their scores of estimated losses. Programmes with 20 per cent or more loss due to default and interest rate differential were classified as unsustainable. Programmes with 10 per cent of loss or more, but less than 20, were classified as potentially sustainable (low), and those with between 2 and 10 per cent loss as potentially sustainable (high). In other words, these programmes are assumed to be able to become sustainable if adequate measures are taken to improve financial performance. However, in the case of a potentially sustainable (low) programmes, reforms will need to be very substantial. Finally, programmes with less than two per cent loss due to default and interest rate differential were classified as sustainable.

The procedure yielded four cases qualified as unsustainable, five cases as potentially sustainable (low), five cases as potentially sustainable (high), and only three cases as sustainable. Two cases were not classified due to insufficient information.[7]

The three cases classified as sustainable programmes included one programme designed as a specialized credit project, one programme administered by an NGO, and one programme administered by a marketing cooperative. In the latter case, credit is provided as part of an interlinked contract contributing greatly to its sustainability.

As a general conclusion we can confirm that, in the present situation, most non-conventional credit programmes of our analysis are not sustainable. Only three programmes were classified as sustainable. The potentially sustainable programmes might become sustainable in the future, but only if reforms are implemented.

In order to analyse the factors that explain sustainability of the programme, scores were assigned to each of the variables discussed above, that is (1) *screening*, (2) *supervision*, (3) *enforcement*, (4) *professional capacity*, (5) *separation of functions*, (6) *ownership*, (7) *scale of operations*, (8) *operating cost, and* (9) *interest rate policy*. Each variable can be positive (+), intermediate (±), or negative (−). The scores are shown in Table 13.2.

In the next step of the analysis average values were calculated for each group of programmes and each explaining variable. Positive values were rated at 10, intermediate values at 5 and negative values at 0. Average values can take any value between 0 (when all cases in the group were assigned a negative value) and 10 (all cases a positive value).

Results for average scores are shown in Table 13.3. The most significant differences between scores in Group 1 (unsustainable programmes) and Group 4 (sustainable programmes) occur for the variables *screening, enforcement, separation of functions, operating cost* and *interest rate policy. Operating cost* and *interest rate policy* are highly related in the range of more sustainable programmes, as the score for *interest rate policy* cannot obtain a positive value if interest rates do not cover operating costs. So the variables *screening, enforcement, separation of functions* and *interest rate policy* can be considered the key variables which determine the differences between unsustainable and sustainable programmes.

Some other variables, *supervision, ownership*, and *scale of operations* show smaller differences between the groups. *Supervision* of credit turns out to be less of a determinant variable. Rural development projects often argue that intensive supervision is necessary, but at the same time cost of

Table 13.2   Estimated loss due to default on repayments and interest rate policies, and scores assigned for explaining variables[1] for 18 non-conventional credit programmes in Central America

	Loss	*Scr.*	*Sup.*	*Enf.*	*Pro. cap.*	*Sep. fun.*	*Own*	*Sca.*	*Ope. cos.*	*Int. rat.*
Group 1										
NGO 1	37%	–	–	–	–	–	+	–	–	–
NGO 5	28%	–	±	–	–	–	+	–	–	–
NGO 3	24%	–	±	–	–	–	+	–	–	–
COOP 5	21%	–	+	–	±	–	+	+	+	±
Group 2										
RDP 5	18%	–	+	±	+	+	–	+	±	–
RDP 2	15%	+	±	+	±	±	–	+	–	–
RDP 1	15%	+	+	±	+	+	–	+	–	–
COOP 1	11%	–	+	–	–	–	+	–	–	±
RDP 4	10%	–	+	±	+	±	–	+	±	–
Group 3										
COOP 2	8%	±	+	–	–	–	+	–	–	±
COOP 4	8%	+	+	+	–	±	±	–	–	–
CRP 2	8%	+	+	+	+	±	+	±	±	±
CRP 3	6%	+	+	–	±	±	+	±	+	±
NGO 4[2]	6%	+	+/±	+/±	±	±	+	–	–	±
Group 4										
COOP 3	1%	+	+	+	–	±	+	–	+	±
NGO 6	0%	+	+	+	+	+	+	+	+	+
CRP 1	–2%	+	+	+	+	+	–	±	±	+

[1] Explaining variables are: screening (Scr.), supervision (Sup.), enforcement (Enf.), professional capacity (Pro.cap.), separation of functions (Sep.fun.), ownership (Own.), scale of operations (Sca.), operating cost (Ope.cos.) and interest rate policy (Int.rat.).
[2] As this programme works with two rather different subprograms, two scores are shown for supervision and enforcement.
Explanation of groups:
Group 1: Unsustainable; Group 2: Potentially sustainable (low); Group 3: Potentially sustainable (high); Group 4: Sustainable.

supervision will increment, and apparently in many cases studied, intensive supervision did not contribute enough to maintain losses at acceptable levels. In reality, fungibility of credit makes it rather difficult for supervision to be really effective. An even more important explanation for the

Table 13.3   Averages of estimated loss due to default on repayments and interest rate policies, and of scores assigned for explaining variables[1] for groups of non-conventional credit programmes in Central America[2]

Group	Loss	Scr.	Supl.	Enf.	Pro. cap	Sep fun.	Own	Scale	Ope. cost	Int. rate
1	20% or more	0	5	0	1	0	10	2.5	2.5	1
2	10%–< 20%	4	9	5	7	6	2	8	2	1
3	2%–< 10%	9	9.5	5.5	4	4	9	2	3	4
4	less than 2%	10	10	10	7	8	7	5	8	8

[1] see note to Table 13.2.
[2] Group 1: unsustainable programmes; Group 2: Potentially sustainable (low) programs; Group 3: Potentially sustainable (high) programmes; Group 4: Sustainable programmes.

lack of the relation in our sample may be that many programmes (especially those with intensive technical supervision, like rural development projects) offer credit in order to promote a transformation of production activities, for example, the introduction of new crops and technologies. Most often, credit is used for high risk activities. Supervision may well be interpreted as a – partial – compensation for restricting the use of credit for high risk investments.

In the case of *ownership* an ambiguous relationship was found. While it might be expected that a clear sense of ownership of funds, or the financial obligation to reimburse the loans which finance credit operations (none of the programmes relied substantially on saving deposits for financing credit operations) will be an incentive for the implementing agency to manage the programme in a sustainable manner, they do not behave in this way in our case studies. Several factors may play a role in explaining this contrary finding. One aspect may be that donated funds are not really felt as own funds, and do not offer an incentive to maintain the value of the fund. On the contrary, in several cases there appear to be expectations that the probability of future donations of credit funds is higher if there is a lack of funds, than if former loans are recovered. For intermediating organizations it is easier to gain surpluses on new funds than on recycled funds. Another factor may be the use of credit funds to maintain patron–client relationships between the intermediating organization and borrowers. If the policies of the non-conventional credit programme are soft, this will enhance social or political support for the programme.

Scale of operations does not appear to be a main determinant variable, due to the fact that small credit funds operated by marketing institutions can be rather efficient, and achieve sustainability in spite of their small scale. On the opposite side, the larger funds appear to show scale disadvantages, as they have to confront a limited absorption capacity in the target population, which induces them to supply credit to households which lack sufficient investment capacity.

The variable *professional capacity* shows a more irregular pattern. One might expect that professional capacity to be positively related to sustainability of the programme. However, two factors play a role with opposite effects. First, several programmes are administered as trust funds by banks or credit and savings cooperatives, but do not show sustainability. Notably, some of them incur losses because of low interest rates. In our case studies this was due to the financial institution and not to policies imposed by the programme. On the other hand, administration of a trust fund by a financial institution does not guarantee adequate screening, as this institution is not exposed to financial risks. On the contrary, perverse incentives exist in the cases where administration fees depend on disbursements or credit portfolio. Second, the marketing cooperatives administrating credit funds do not have professional capacity for this activity but still manage to be classed in the higher groups.

The key variables for sustainability of the programme appear to interfere in some logical sequence or hierarchy. All key variables have very low values in the unsustainable programmes. The main determinant for differences between the unsustainable and potential sustainable (low) programmes can be identified as being the *separation of functions* and, in a somewhat lesser degree, *enforcement* variables. When comparing the next two groups the main determinant for the difference between them appears to be the *screening* variable. Finally, when comparing the potentially sustainable (high) and sustainable programmes, the difference is mainly determined by the variables *enforcement* and *interest rate policy*.[8]

The unsustainable programmes suffer problems in all areas represented by the key variables. This is mainly due to an inadequate institutional design. The lack of separation of functions prohibits appropriate screening, as well as enforcement of repayment, leading to very high default rates, and thus large financial losses. When a more adequate institutional design is applied, screening can be improved, as well as enforcement. However, many of these programmes still lack an appropriate interest rate policy. Self imposed interest rate restrictions contribute to a lack of financial sustainability, very similar to the former experiences of national development banks (who suffered restrictions, often imposed by governmental directives). Moreover, in many programmes the willingness to apply appropriate enforcement mechanisms often still remains insufficient. Sometimes, this may also be attributed to problems of institutional design because the responsibility for recovery has to be taken by persons or institutions (projects) who do not want to tarnish their image as they are also responsible for promotional functions. Both factors (self imposed interest rate restrictions and lack of application of appropriate enforcement mechanisms) contribute to unsustainability of the programme.

In the analysis we only looked at credit as an isolated financial service offered by non-conventional credit programmes. In fact, most of these programmes do not engage in facilitating services for savings deposits or other financial services. There are a few exceptions, mainly those programmes that promote community banks (included in two of the 19 case studies). Apart from the lack of intention to attract saving deposits, which could contribute to sustainability of the programme, the legal framework in the Central American countries does not allow for this function, as mentioned before. Even if both conditions were satisfied, self imposed interest rate restrictions might reduce the feasibility of attracting deposits.

4 CONCLUSIONS

A comparison between the determinants of the unsustainability of non-conventional credit programmes in Central America and the results of former interventions in the rural credit market by the establishment of agricultural development banks, reveals many similarities and a few distinct features.

The criticism of the agricultural development banks that interest rate restrictions caused regressive income distribution effects does not hold for non-conventional credit programmes. The latter have not attracted large farmers in Central America, mainly because loans are smaller and because of intensive monitoring by project staff and donor organizations.

The other criticism of agricultural development banks, that is lack of financial sustainability, is repeated by most non-conventional programmes. In these programmes, interest rate restrictions (in this case mainly self imposed), high default rates, ineffective screening, and high administration costs inhibit financial sustainability, as was the case with the agricultural development banks. Weak enforcement procedures and inadequate screening may even be worse than in development banks, mainly as a result of an inappropriate institutional design. A mix of functions of promotion, credit approval and credit recovery is an obstacle to effective screening and enforcement in several programmes.

The main advantage of non-conventional credit programmes over development banks is their greater flexibility to adapt their schemes to the target group. Development banks lack incentives for institutional innovations, while non-conventional credit programmes do have possibilities for these innovations and there are incentives for them, at least at the overall programme level. Those programmes that manage to introduce effective innovations, for example in the area of recovery strategies, and succeed in obtaining both improved access to poor rural households and better financial performance than other programmes, are rewarded by donor organizations as they may receive new donations or concessional credits. In our study, the NGO classed as a sustainable programme is a case in point, which rapidly diversified financing sources.

The greater flexibility for operational implementation of non-conventional programmes also determines the great diversity in the programmes we studied. So, even if we can confirm the general conclusion that most non-conventional credit programmes repeat previous errors and failures of agricultural development banks, there are exceptions and opportunities for change.

Notes

1. The effectiveness of this mechanism depends upon the existence of assets or income that supports the co-debtor's position. In the case investigated, there is evidence that the cooperative federations that sign as co-debtors for their cooperatives do not possess enough capital to honour that commitment. This will probably cause problems with the enforcement of the repayment.
2. A peculiar variant of this dilution of responsibilities is the legal rule that the individual members should not be held responsible for the debts of collective investment projects, and that the loans of collective production should be paid for by the income generated collectively. This interpretation is based on an interpretation of the concept of limited responsibility, as included in co-operative legislation. The ultimate consequence is that in case of default, the only way to recover part of the loan is to present bankruptcy of the cooperative, which is seldom done.
3. Interlinkages with the land and labour market can be found among informal private lenders (for example moneylenders, big farmer lenders).
4. Bastiaensen (1995) calls this the 'progressivity principle', following the terminology of the Nitlapán institute.
5. It is significant that credit unions in Honduras, which used to accept movable collateral, have completely eliminated this type of collateral, because of the difficulties involved in enforcement.
6. This is one of the reasons why a series of financial reforms, that should enhance competitiveness in the financial sector and reduce the burden of state debt on the financial market is now being promoted.

7. One of these cases is the peasant shop case. In this case no interest rates are charged, just as competitors do. However, marketing margins should be sufficiently high to cover credit cost, which was definitely not true for our case study. However, we do not have data to quantify losses to be assigned to the credit programme. The other programme had started too recently to generate information on default.
8. Looking at these groups, only the variable *separation of functions* shows values equal to the variable *interest rate policies*. However, the value of the former showed some irregularity in Group 3, if we also look at its value in Group 2.

Bibliography

Adams, D.W., D.H. Graham and J.D. Von Pischke (eds) (1984) *Undermining Rural Development with Cheap Credit* (Boulder: Westview Press).

Bastiaensen, J. (1995) 'Institution building for equitable rural development in Nicaragua: a view on the strategic role of non-conventional finance', paper presented at the X ASERCCA Annual Conference, Paris.

Banco Central de Nicaragua (1994) *El crédito en Nicaragua* (processed).

Benito, C. (1993) *Debt Overhang and Other Barriers to the Growth of Agriculture in El Salvador*, Agricultural Policy Analysis Project for US-AID/El Salvador (processed).

Braverman, A. and J.L. Guasch (1993) 'Administrative failures in rural credit programmes', in K. Hoff *et al.*, *The economics of rural organization* (Washington: The World Bank).

Cuevas, C.E. *et al.* (1991) *El sector financiero informal en El Salvador*, Report to US-AID El Salvador, Ohio State University and FUSADES.

González Vega, C. *et al.* (1993) *Financiamiento de la microempresa rural: FINCA Costa Rica*, Ohio State University/Academia de Centroamérica.

González Vega, C. and J.I. Torrico (1995) *Honduras: mercados financieros rurales no formales*, Proyecto para el Desarrollo de Políticas Agfrícolas de Honduras.

Hulme, D. (1995) 'Solving agrarian questions through finance? Financial innovations, rural poverty and vulnerability', in *Agrarian Questions: the politics of farming anno 1995; proceedings*, Vol. II, pp. 647–65, Wageningen: Wageningen Agricultural University.

Nitlapán-UCA (1994) *Situación y perspectivas de las nuevas estructuras institucionales de financiamiento rural – informe principal*, report to SIDA.

Pischke, J.D. Von (1991) *Finance at the Frontier. Debt capacity and the role of credit in the private economy*, EDI Development Studies (Washington, DC: The World Bank).

Quintanilla, J.D. (1995) *Sistemas de financiamiento rural en El Salvador*, preliminary report to RUTA/- World Bank (processed).

Stiglitz, J.E. (1990) 'Peer monitoring and credit markets', *World Bank Economic Review*, Vol. 4, nr.3, pp. 351–66.

Wenner, M. (1995) 'Group credit: a means to improve information transfer and loan repayment performance', *Journal of Development Studies*, Vol. 32, No. 2, pp. 263–81.

Index